From "Antinomy"—

How long has it been since I went to bed "last night"?

The view from the hospital room was only a little startling—but Virginia had no recollection of having entered any hospital, or even of having been ill. And the sight of the policeman flying above the rooftops in an oversized garbage can left no room for doubt that she had awakened a *long* time after the last time she remembered going to sleep.

Let's see. Time travel, huh? That means . . .

The door opened to admit a young man.

"What year is this, anyway?" she asked; he opened his mouth, then shut it.

"And what did I die of?"

THE LATEST SCIENCE FICTION FROM DELL BOOKS

Also by Spider Robinson
in Dell Edition
STARDANCE
(written with Jeanne Robinson)

*denotes an illustrated book

ANTINOMY

Spider Robinson

A DELL BOOK

this one is for
Jim Frenkel, of course

Published by
Dell Publishing Co., Inc.
1 Dag Hammarskjold Plaza
New York, New York 10017

"Antinomy" first appeared in *Destinies*; "Half an Oaf" in *Analog Annual 1976*; "Too Soon We Grow Old" in *Analog Yearbook*; "When No Man Pursueth" in *Analog*; "Nobody Likes to be Lonely," "No Renewal," and "Overdose" in *Galaxy*; "Satan's Children" in *New Voices*; "Apogee" in *Borealis*; "Tin Ear" in *Cosmos*; and "The Magnificent Conspiracy" in *Chrysalis 1*.

Dell ® TM 681510, Dell Publishing Co., Inc.

ISBN: 0-440-10235-9

Printed in the United States of America

Cover painting by Larry Kresek, based on photo studies by Jay Kay Klein

First printing—October 1980

CONTENTS

ANTINOMY

a collection of things
written, illustrated, scored and sung
by Spider Robinson

INTRODUCTION:
Welcome to the Antinomy Mine

I said antinomy, not antimony.

Antimony is a metallic element whose most common allotropic form is a hard, brittle, lustrous, silver-white crystalline material. It is used in a variety of alloys, with lead in battery plates, for example, and is also used for flameproofing, paints, semiconductors, and ceramics. It has nothing whatsoever to do with this collection.

Antinomy is, essentially, a state of mind. I don't propose to define it here, as that is done by one of the characters in the title story, and I refuse to steal her thunder. But perhaps I could say that antinomy is the state of mind in which the soul seems a hard, brittle, silver-white crystalline material, carrying enormous current and beset by hellfire. Antinomy is the very stuff of fiction, the very stuff of art, and it is that substance which I mine for a living.

I cannot show you the actual mine itself. It is located somewhere within my skull (or at least has a branch office there), and there's only room in there for one of us. But I can show you the warehouse. It is a battered cardboard box, which sits at the moment behind me on my office floor. On one end a gummed label has been slapped over the faded original legend ("Gordon's Gin"), identifying the box's contents as *Pay Copy—Sold.*

It contains my lifework to date, and you and I are about to rummage through it together.

Well, just some of the high spots—and not even all

of *them*. There's nearly half a million words in there. We'll skip the novels, of course, and we'll pass over the ones that were written by mistake and pseudonymously published to clothe the baby, and we'll omit anything that is presently available elsewhere in paperback—even though that cuts out anything that has ever won me an award.

What we have left is eleven stories that I am really proud of and believe you will enjoy.

Plus a few tidbits.

It was, I swear, Jim Frenkel's idea. The proposal I submitted to him at Dell for this book contained stories and nothing but stories, so help me. He bought it on that basis. And then he started thinking about all the money he was paying me just to type up a table of contents, and he came to see me.

"You're more than just a writer, Spider," he said to me, lighting my cigarette. "You play the guitar and sing with Jeanne, you've written what, fifty songs, you're all the time sketching and cartooning, you even" (here, to give him credit, he shuddered) "create puns. Now what I envision for this collection is a kind of Book of Spider Robinson, a really representative selection of all the things you do. Song lyrics, I'll have somebody draw up sheet music, cartoons, essays, puns, you name it. Introductions to each story explaining how it came to be written, nice tidbits like that. What do you think?"

I can produce in evidence a photograph of the Indian Rope Burn he gave me twisting my arm.

I have modified his plan somewhat. Whenever I read a collection with story intros, I invariably read the story first and then the intro. I find it faintly frustrating to be told about a story I haven't read yet. Consequently I have, with one exception, placed discussion periods, if any, *after* the story.

Even this caused me some trepidation, for I truly do believe the maxim that if a story really *needs* anything said *about* it, then it has failed as a story. But whenever someone writes or comes up to me at a convention to discuss a story of mine, what they almost always want to hear is *anecdotes about* the stories: how I "got the idea" or how long it took to sell it or where I was living at the time. I am forced to conclude that the science fiction readership as a group has an enormous interest in the process of writing science fiction itself. Consequently I have placed here and there (whenever there was an anecdote that seemed worth the telling) three afterwords, an interleaf between two related stories, and one foreword.

There are also a few cartoons, four songs (two of them illustrated), an illustration for material of mine that does not appear in this book, and—be thou warned—several reasonably dreadful puns (those who work in antinomy mines are subject to a ghastly disease called Black Brain).

You are perfectly welcome to ignore any or all of the tidbits. They are only meant as a kind of icing on the cake, and a cake that can't make it without icing doesn't belong in your dining room.

But don't ignore the stories. The guitar, the songs, the doodles chasing each other across all my first drafts, the happy chatter, all of that is stuff that I *do*. What I *am* is a writer.* It has cost me to write these stories. Some of them are "only" entertainments, especially the early ones. They are the best entertainments I could construct at the time, and I like them in retrospect. Others carry more freight than that, especially the later ones. They are the best justifications I am able to offer for having engaged your attention in the first place. The beginning, middle and end stories— "Antinomy," "Satan's Children," and "The Magnificent Conspiracy"—are the last three things that have emerged from my typewriter (two of them have not yet seen publication anywhere as I write this), and

---

*—and a husband and father, but those are different stories.

one of them, "When No Man Pursueth," is the fifth or sixth story I ever wrote.

They were all hauled up from the antinomy mine at great expense of effort; I hope you enjoy them.

Spider Robinson
Halifax, Nova Scotia
October, 1978

1
ANTINOMY

The first awakening was just awful.

She was naked and terribly cold. She appeared to be in a plastic coffin, from whose walls grew wrinkled plastic arms with plastic hands that did things to her. Most of the things hurt dreadfully. *But I don't have nightmares like this,* she thought wildly. She tried to say it aloud, and it came out, "A."

Even allowing for the sound-deadening coffin walls, the voice sounded distant. "Christ, she's awake already."

Eyes appeared over hers, through a transparent panel she had failed to see since it had showed only a ceiling the same color as the coffin's interior. The face was masked and capped in white, the eyes pouched in wrinkles. *Marcus Welby. Now it makes enough sense. Now I'll believe it. I don't have nightmares like this.*

"I believe you're right." The voice was professionally detached. A plastic hand selected something that lay by her side, pressed it to her arm. "There."

Thank you, Doctor. If my brain doesn't want to remember what you're operating on me for, I don't much suppose it'll want to record the operation itself. Bye.

She slept.

The second awakening was better.

She was astonished not to hurt. She had expected to hurt, somewhere, although she had also expected to be too dopey to pay it any mind. Neither condition obtained.

She was definitely in a hospital, although some of the gadgetry seemed absurdly ultramodern. *This certainly isn't Bellevue,* she mused. *I must have contracted something fancy. How long has it been since I went to bed "last night"?*

Her hands were folded across her belly; her right hand held something hard. It turned out to be a traditional nurse-call buzzer—save that it was cordless. Lifting her arm to examine it had told her how terribly weak she was, but she thumbed the button easily—it was not spring-loaded. "Nice hospital," she said aloud, and her voice sounded too high. *Something with my throat? Or my ears? Or my . . . brain?*

The buzzer might be improved, but the other end of the process had not changed appreciably; no one appeared for a while. She awarded her attention to the window beside her, no contest in a hospital room, and what she saw through it startled her profoundly.

She *was* in Bellevue, after all, rather high up in the new tower: the rooftops below her across the street and the river beyond them told her that. But she absorbed the datum almost unconsciously, much more startled by the policeman who was flying above those rooftops, a few hundred feet away, in an oversize garbage can.

Yep, my brain. The operation was a failure, but the patient lived.

For a ghastly moment there was great abyss within her, into which she must surely fall. But her mind had more strength than her body. She willed the abyss to disappear, and it did. *I may be insane, but I'm not going to go nuts over it,* she thought, and giggled. She decided the giggle was a healthy sign, and did it again, realizing her error when she found she could not stop.

It was mercifully shorter than such episodes usually are; she simply lost the strength to giggle. The room swam for a while, then, but lucidity returned rather rapidly.

Let's see. Time travel, huh? That means . . .

The door opened to admit—not a nurse—but a young man of about twenty-five, five years her junior. He was tall and somehow self-effacing. His clothes and appearance did not strike her as conservative, but she decided they probably were—for this era. He did not look like a man who would preen more than convention required. He wore a sidearm, but his hand was nowhere near the grip.

"What year is this, anyway?" she asked as he opened his mouth, and he closed it. He began to look elated and opened his mouth again, and she said "And what did I die of?" and he closed it again. He was silent then for a moment, and when he had worked it out she could see that the elation was gone.

But in its place was a subtler, more personal pleasure. "I congratulate you on the speed of your uptake," he said pleasantly. "You've just saved me most of twenty minutes of hard work."

"The hell you say. I can deduce what *happened*, all right, but that saves you twenty seconds, max. 'How' and 'why' are going to take just as long as you expected. And don't forget 'when.'" Her voice still seemed too high, though less so.

"How about 'who'? I'm Bill McLaughlin."

"I'm Marie Antoinette, *what the hell year is it?*" The italics cost her the last of her energy; as he replied "1995," his voice faded and the phosphor dots of her vision began to enlarge and drift apart. She was too bemused by his answer to be annoyed.

Something happened to her arm again, and picture and sound returned with even greater clarity. "Forgive me, Ms. Harding. The first thing I'm supposed to do is give you the stimulant. But then the first thing you're supposed to do is be semiconscious."

"And we've dispensed with the second thing," she said, her voice normal again now, "which is telling me that I've been a corpsicle for ten years. So tell me why, and why I don't *remember* any of it. As far as I know I went to sleep last night and woke up here, with a brief interlude inside something that must have been a defroster."

"I thought you *had* remembered, from your first question. I hoped you had, Ms. Harding. You'd have been the first . . . never mind—your next question made it plain that you don't. Very briefly, ten years ago you discovered that you had leukemia . . ."

"Myelocytic or lymphocytic?"

"Neither. Acute."

She paled. "No wonder I've suppressed the memory."

"You haven't. Let me finish. Acute Luke was the diagnosis, a new rogue variant with a bitch's bastard of a prognosis. In a little under sixteen weeks they tried corticosteroids, L-aspiraginase, cytosine arabinoside, massive irradiation, and mercrystate crystals, with no more success than they'd expected, which was none and negatory. They told you that the new bone-marrow transplant idea showed great promise, but it might be a few years. And so you elected to become a corpsicle. You took another few weeks arranging your affairs and then went to a Cold Sleep Center and had yourself frozen."

"*Alive?*"

"They had just announced the big breakthrough. A week of drugs and a high-helium atmosphere and you can defrost a living person instead of preserved meat. You got in on the ground floor."

"And the catch?"

"The process scrubs the top six months to a year off your memory."

"Why?"

"I've been throwing around terminology to demonstrate how thoroughly I've read your file. But I'm not a doctor. I don't understand the alleged 'explanation' they gave me, and I dare say you won't either."

"Okay." She forgot the matter, instantly and forever. "If you're not a doctor, who are you, Mr. McLaughlin?"

"Bill. I'm an Orientator. The phrase won't be familiar to you—"

"—but I can figure it out, Bill. Unless things have slowed down considerably since I was alive, ten years is a hell of a jump. You're going to teach me how to dress and speak and recognize the ladies' room."

"And hopefully to stay alive."

"For how long? Did they fix it?"

"Yes. A spinal implant, right after you were thawed. It releases a white-cell antagonist into your bloodstream, and it's triggered by a white-cell surplus. The antagonist favors rogue cells."

"Slick. I always liked feedback control. Is it foolproof?"

"Is anything? Oh, you'll need a new implant every five years, and you'll have to take a week of chemotherapy here to make sure the implant isn't rejected before we can let you go. But the worst side-effect we know of is partial hair-loss. You're fixed, Ms. Harding."

She relaxed all over, for the first time since the start of the conversation. With the relaxation came a dreamy feeling, and she knew she had been subtly drugged, and was pleased that she had resisted it, quite unconsciously, for as long as had been necessary. She disliked don't-worry drugs; she preferred to worry if she had a mind to.

"Virginia. Not Ms. Harding. And I'm pleased with the Orientator I drew, Bill. It will take you awhile to get to the nut, but you haven't said a single inane thing yet, which under the circumstances makes you a remarkable person."

"I like to think so, Virginia. By the way, you'll doubtless be pleased to know that your fortune has come through the last ten years intact. In fact, it's actually grown considerably."

"There goes your no-hitter."

"Beg pardon?"

"Two stupid statements in one breath. First, of *course* my fortune has grown. A fortune the size of mine can't *help* but grow—which is one of the major faults of our economic system. What could be sillier than a goose that insists on burying you in golden eggs? Which leads to number two: I'm anything but pleased. I was hoping against hope that I was broke."

His face worked briefly, ending in a puzzled frown. "You're probably right on the first count, but I think the second is ignorance rather than stupidity. I've never been rich." His tone was almost wistful.

"Count your blessings. And be grateful you can count that high."

He looked dubious. "I suppose I'll have to take your word for it."

"When do I start getting hungry?"

"Tomorrow. You can walk now, if you don't overdo it, and in about an hour you'll be required to sleep."

"Well, let's go."

"Where to?"

"Eh? *Outside*, Bill. Or the nearest balcony or solarium. I haven't had a breath of fresh air in ten years."

"The solarium it is."

As he was helping her into a robe and slippers the door chimed and opened again, admitting a man in the time-honored white garb of a medical man on duty, save that the stethoscope around his neck was as cordless as the call-buzzer had been. The pickup was doubtless in his breast pocket, and she was willing to bet that it was warm to the skin.

The newcomer appeared to be a few years older than she, a pleasant-looking man with gray-ribbed temples and plain features. She recognized the wrinkled eyes and knew he was the doctor who had peered into her plastic coffin.

McLaughlin said, "Hello, Dr. Higgins. Virginia

Harding, Dr. Thomas Higgins, Bellevue's Director of Cryonics."

Higgins met her eyes squarely and bowed. "Ms. Harding. I'm pleased to see you up and about."

Still has the same detached voice. Stuffy man. "You did a good job on me, Dr. Higgins."

"Except for a moment of premature consciousness, yes, I did. But the machines say you weren't harmed psychologically, and I'm inclined to believe them."

"They're right. I'm some tough."

"I know. That's why I brought you up to Level One Awareness in a half-day instead of a week. I knew your subconscious would fret less."

Discriminating machines, she thought. *I don't know that I like that.*

"Doctor," McLaughlin cut in, "I hate to cut you off, but Ms. Harding has asked for fresh air, and—"

"—and has less than an hour of consciousness left today. I understand. Don't let me keep you."

"Thank you, Doctor," Virginia Harding said. "I'd like to speak further with you tomorrow, if you're free."

He almost frowned, caught himself. "Later in the week, perhaps. Enjoy your walk."

"I shall. Oh, how I shall. Thank you again."

"Thank Hoskins and Parvati. They did the implant."

"I will, tomorrow. Good-bye, Doctor."

She left with McLaughlin, and as soon as the door had closed behind them, Higgins went to the window and slammed his fist into it squarely, shattering the shatterproof glass and two knuckles. Shards dropped thirty long stories, and he did not hear them land.

McLaughlin entered the office and closed the door.

Higgins's office was not spare or austere. The furnishings were many and comfortable, and in fact the entire room had a lived-in air which hinted that Higgins's apartment might well be spare and austere.

Shelves of books covered two walls; most looked medical and all looked used. The predominant color of the room was black—not at all a fashionable color—but in no single instance was the black morbid, any more than is the night sky. It gave a special vividness to the flowers on the desk, which were the red of rubies, and to the profusion of hand-tended plants which sat beneath the broad east window (now opaqued) in a riotous splash of many colors for which our language has only the single word "green." It put crisper outlines on anything that moved in the office, brought both visitors and owner into sharper relief.

But the owner was not making use of this sharpening of perception at the moment. He was staring fixedly down at his desk; precisely, in fact, at the empty place where a man will put a picture of his wife and family if he has them. He could not have seen McLaughlin if he tried; his eyes were blinded with tears. Had McLaughlin not seen them, he might have thought the other to be in an autohypnotic trance or a warm creative fog, neither of which states were unusual enough to call for comment.

Since he did, he did not back silently out of the office. "Tom." There was no response. "Tom," he said again, a little louder, and then "TOM!"

"Yes?" Higgins said evenly, sounding like a man talking on an intercom. His gaze remained fixed, but the deep-set wrinkles around it relaxed a bit.

"She's asleep."

Higgins nodded. He took a bottle from an open drawer and swallowed long. He didn't have to uncap it first, and there weren't many swallows that size left. He set it, clumsily, on the desk.

"For God's sake, Tom," McLaughlin said half-angrily. "You remind me of Monsieur Rick in *Casablanca*. Want me to play 'As Time Goes By' now?"

Higgins looked up for the first time, and smiled beatifically. "You might," he said, voice steady. " 'You *must* remember this . . . as time goes by.' " He smiled again. "I often wonder." He looked down again, obviously forgetting McLaughlin's existence.

Self-pity in this man shocked McLaughlin, and cheerful self-pity disturbed him profoundly. "Jesus," he said harshly. "That bad?" Higgins did not hear. He saw Higgins's hand then, with its half-glove of bandage, and sucked air through his teeth. He called Higgins's name again, elicited no reaction at all.

He sighed, drew his gun and put a slug into the ceiling. The roar filled the office, trapped by sound-proofing. Higgins started violently, becoming fully aware just as his own gun cleared the holster. He seemed quite sober.

"Now that I've got your attention," McLaughlin said dryly, "would you care to tell me about it?"

"No." Higgins grimaced. "Yes and no. I don't suppose I have much choice. She didn't remember a thing." His voice changed for the last sentence; it was very nearly a question.

"No, she didn't."

"None of them have yet. Almost a hundred awakenings, and not one remembers anything that happened more than ten to twelve months before they were put to sleep. And still somehow I hoped . . . I had hope . . ."

McLaughlin's voice was firm. "When you gave me her file, you said you 'used to know her,' and that you didn't want to go near her 'to avoid upsetting her.' You asked me to give her special attention, to take the best possible care of her, and you threw in some flattery about me being your best Orientator. Then you come barging into her room on no pretext at all, chat aimlessly, break your hand and get drunk. So you loved her. And you loved her in her last year."

"I diagnosed her leukemia," Higgins said emotionlessly. "It's hard to miss upper abdomen swelling and lymph node swelling in the groin when you're making love, but I managed for weeks. It was after she had the tooth pulled and it wouldn't stop bleeding that . . ." He trailed off.

"She loved you too."

"Yes." Higgins's voice was bleak, hollow.

"Bleeding Christ, Tom," McLaughlin burst out. "Couldn't you have waited to . . ." He broke off, thinking bitterly that Virginia Harding had given him too much credit.

"We tried to. We knew that every day we waited decreased her chances of surviving cryology, but we tried. She insisted that we try. Then the crisis came . . . oh damn it, Bill, *damn* it."

McLaughlin was glad to hear the profanity—it was the first sign of steam blowing off. "Well, she's alive and healthy now."

"Yes. I've been thanking God for that for three months now, ever since Hoskins and Parvati announced the unequivocal success of spinal implants. I've thanked God over ten thousand times, and I don't think He believed me once. I don't think *I* believed me once. Now doesn't that make me a selfish son of a bitch?"

McLaughlin grinned. "Head of Department and you live like a monk, because you're selfish. For years, every dime you make disappears down a hole somewhere, and everybody wonders why you're so friendly with Hoskins & Parvati, who aren't even in your own *department*, and only now, as I'm figuring out where the money's been going, do I realize what a truly selfish son of a bitch you are, Higgins."

Higgins smiled horribly. "We talked about it a lot, that last month. I wanted to be frozen too, for as long as they had to freeze her."

"What would that have accomplished? Then neither of you would have remembered."

"But we'd have entered and left freeze at the *same time*, and come out of it with sets of memories that ran nearly to the day we met. We'd effectively be precisely the people who fell in love once before; we could have left notes for ourselves and the rest would've been inevitable. But she wouldn't hear of it. She pointed out that the period in question could be any fraction of forever, with no warranty. I insisted, and got quite histrionic about it. Finally she brought up our age difference."

"I wondered about the chronology."

"She was thirty, I was twenty-five. Your age. It was something we kidded about, but it stung a bit when we did. So she asked me to wait five years, and then if I still wanted to be frozen, fine. In those five years I clawed my way up to head of section here, because I wanted to do everything I could to ensure her survival. And in the fifth year they thought her type of leukemia might be curable with marrow transplants, so I hung around for the two years it took to be sure they were wrong. And in the eighth year Hoskins started looking for a safe white-cell antagonist, and again I had to stay room temperature to finance him, because nobody else could smell that he was a genius. When he met Parvati, I knew they'd lick it, and I told myself that if they needed me, that meant she needed me. I wasn't wealthy like her—I had to keep working to keep them both funded properly. So I stayed."

Higgins rubbed his eyes, then made his hands lie very still before him, left on right. "Now there's a ten-year span between us, the more pronounced because she hasn't experienced a single minute of it. Will she love me again or won't she?" The bandaged right hand escaped from the left, began to tap on the desk. "For ten years I told myself I could stand to know the answer to that question. For ten years it was the last thing I thought before I fell asleep and the first thing I thought when I woke up. *Will she love me or won't she?*

"She made me promise that I'd tell her everything when she was awakened, that I'd tell her how our love had been. She swore that she'd love me again. I promised, and she must have known I lied, or suspected it, because she left a ten-page letter to herself in her file. The day I became Department Head I burned the fucking thing. I don't want her to love me because she thinks she should.

"Will she love me or won't she? For ten years I believed I could face the answer. Then it came time to wake her up, and I lost my nerve. I couldn't stand to know the answer. I gave her file to you.

"And then I saw her on the monitor, heard her voice coming out of my desk, and I knew I couldn't stand *not* to know."

He reached clumsily for the bottle, and knocked it clear off the desk. Incredibly, it contrived to shatter on the thick black carpet, staining it a deeper black. He considered this, while the autovac cleaned up the glass, clacking in disapproval.

"Do you know a liquor store that delivers?"

"In *this* day and age?" McLaughlin exclaimed, but Higgins was not listening. "Jesus Christ," he said suddenly. "Here." He produced a flask and passed it across the desk.

Higgins looked him in the eye. "Thanks, Bill." He drank.

McLaughlin took a long swallow himself and passed it back. They sat in silence for a while, in a communion and a comradeship as ancient as alcohol, as pain itself. Synthetic leather creaked convincingly as they passed the flask. Their breathing slowed.

If a clock whirs on a deskface and no one is listening, is there really a sound? In a soundproof office with opaqued windows, is it not always night? The two men shared the long night of the present, forsaking past and future, for nearly half an hour, while all around them hundreds upon hundreds worked, wept, smiled, dozed, watched television, screamed, were visited by relatives and friends, smoked, ate, died.

At last McLaughlin sighed and studied his hands. "When I was a grad student," he said to them, "I did a hitch on an Amerind reservation in New Mexico. Got friendly with an old man named Wanoma, face like a map of the desert. Grandfather-grandson relationship—close in that culture. He let me see his own grandfather's bones. He taught me how to pray. One night the son of a nephew, a boy he'd had hopes for, got alone-drunk and fell off a motorcycle. Broke his neck. I heard about it and went to see Wanoma that night. We sat under the moon—it was a harvest moon—and watched a fire until it was ashes. Just after the last coal went dark, Wanoma lifted his head and cried out in Zuni. He cried out, 'Ai-yah, my heart is full of sorrow.'"

McLaughlin glanced up at his boss and took a swallow. "You know, it's impossible for a white man to say those words and not sound silly. Or theatrical. It's a simple statement of a genuine universal, and there's no way for a white man to say it. I've tried two or three times since. You can't say it in English."

Higgins smiled painfully and nodded.

"I cried out too," McLaughlin went on, "after Wanoma did. The English of it was, 'Ai-yah, my brother's heart is full of sorrow. His heart is my heart.' Happens I haven't ever tried to say that since, but you can see it sounds hokey too."

Higgins's smile became less pained, and his eyes lost some of their squint. "Thanks, Bill."

"What'll you do?"

The smile remained. "Whatever I must. I believe I'll take the tour with you day after tomorrow. You can use the extra gun."

The Orientator went poker-faced. "Are you up to it, Tom? You've got to be fair to her, you know."

"I know. Today's world is pretty crazy. She's got a right to integrate herself back into it without tripping over past karma. She'll never know. I'll have control on Thursday, Bill. Partly thanks to you. But you do know why I selected you for her Orientator, don't you?"

"No. I don't think I do."

"I thought you'd at least have suspected. Personality Profiles are a delightful magic. Perhaps if we ever develop a science of psychology we'll understand why we get results out of them. According to the computer, your PP matches almost precisely to my own—of ten years ago. Probably why we get along so well."

"I don't follow."

"Is love a matter of happy accident or a matter of psychological inevitability? Was what 'Ginia and I had fated in the stars, or was it a chance jigsawing of personality traits? Will the woman she was ten years ago love the man I've become? Or the kind of man I was then? Or some third kind?"

"Oh, fine," McLaughlin said, getting angry. "So I'm your competition."

"Aha," Higgins pounced. "You do feel something for her."

"I . . ." McLaughlin got red.

"You're my competition," Higgins said steadily. "And, as you have said, you are my brother. Would you like another drink?"

McLaughlin opened his mouth, then closed it. He rose and left in great haste, and when he had gained the hallway he cannoned into a young nurse with red hair and improbably gray eyes. He mumbled apology and continued on his way, failing to notice her. He did not know Deborah Manning.

Behind him, Higgins passed out.

Throughout the intervening next day Higgins was conscious of eyes on him. He was conscious of little enough else as he sleepwalked through his duties. The immense hospital complex seemed to have been packed full of gray Jello, very near to setting. He plowed doggedly through it, making noises with his mouth, making decisions, making marks on pieces of paper, discharging his responsibilities with the least part of his mind. But he was conscious of the eyes.

A hospital grapevine is like no other on earth. If

you want a message heard by every employee, it is quicker to tell two nurses and an intern than it would be to assemble the staff and make an announcement. Certainly McLaughlin had said nothing, even to his hypothetical closest friend; he knew that any closest friend has at least one *other* closest friend. But at least three OR personnel knew that the Old Man had wakened one personally the other day. And a janitor knew that the Old Man was in the habit of dropping by the vaults once a week or so just after the start of the graveyard shift, to check on the nonexistent progress of a corpsicle named Harding. And the OR team and the janitor worked within the same (admittedly huge) wing, albeit on different floors. So did the clerk-typist in whose purview were Virginia Harding's files, and she was engaged to the anesthetist. Within twenty-four hours, the entire hospital staff and a majority of the patients had added two and two.

(Virginia Harding, of course, heard nary a word, got not so much as a hint. A hospital staff may spill Mercurochrome. It often spills blood. But it never spills beans.)

Eyes watched Higgins all day. And so perhaps it was natural that eyes watched him in his dreams that night. But they did not make him afraid or uneasy. Eyes that watch oneself continuously become, after a time, like a second ego, freeing the first from the burden of introspection. They almost comforted him. They helped.

I have been many places, touched many lives since I touched hers, he thought as he shaved the next morning, *and been changed by them. Will she love me or won't she?*

There were an endless three more hours of work to be taken care of that morning, and then at last the Jell-O dispersed, his vision cleared and she was before him, dressed for the street, chatting with McLaughlin. There were greetings, explanations of some sort were made for his presence in the party, and they left the room, to solve the mouse's maze of corridors that led to the street and the city outside.

It was a warm fall day. The streets were unusually crowded, with people and cars, but he knew they would not seem so to Virginia. The sky seemed unusually overcast, the air particularly muggy, but he knew it would seem otherwise to her. The faces of the pedestrians they passed seemed to him markedly cheerful and optimistic, and he felt that this was a judgment with which she *would* agree. This was not a new pattern of thought for him. For over five years now, since the world she knew had changed enough for him to perceive, he had been accustomed to observe that world in the light of what she would think of it. Having an unconscious standard of comparison, he had marked the changes of the last decade more acutely than his contemporaries, more acutely perhaps than even McLaughlin, whose interest was only professional.

Too, knowing her better than McLaughlin, he was better able to anticipate the questions she would ask. A policeman went overhead in a floater bucket, and McLaughlin began to describe the effects that forcefields were beginning to exert on her transportation holdings and other financial interests. Higgins cut him off before she could, and described the effects single-person flight was having on social and sexual customs, winning a smile from her and a thoughtful look from the Orientator. When McLaughlin began listing some of the unfamiliar gadgetry she could expect to see, Higgins interrupted with a brief sketch of the current state of America's spiritual renaissance. When McLaughlin gave her a personal wrist-phone, Higgins showed her how to set it to refuse calls.

McLaughlin had, of course, already told her a good deal about Civil War Two and the virtual annihilation of the American black, and had been surprised at how little surprised she was. But when, now, he made a passing reference to the unparalleled savagery of the conflict, Higgins saw a chance to make points by partly explaining that bloodiness with a paraphrase of a speech Virginia herself had made ten years before, on the folly of an urban-renewal package concept which had sited low-income housing immediately around urban and suburban transportation hubs. "Built-in disaster," she agreed approvingly, and did not feel obliged to mention that the same thought had occurred to her a decade ago. Higgins permitted himself to be encouraged.

But about that time, as they were approaching one of the new downtown parks, Higgins noticed the expression on McLaughlin's face, and somehow recognized it as one he had seen before—from the inside.

At once he was ashamed of the fatuous pleasure he had been taking in outmaneuvering the younger man. It was a cheap triumph, achieved through unfair advantage. Higgins decided sourly that he would never have forced this "duel with his younger self" unless he had been just this smugly sure of the outcome, and his self-esteem dropped sharply. He shut his mouth and resolved to let McLaughlin lead the conversation.

It immediately took a turning he could not have followed if he tried.

As the trio entered the park, they passed a group of teen-agers. Higgins paid them no mind—he had long since reached the age when adolescents, especially in groups, regarded him as an alien life form, and he was nearly ready to agree with them. But he noticed Virginia Harding noticing them, and followed her gaze.

The group was talking in loud voices, the incomprehensible gibberish of the young. There was nothing Higgins could see about them that Harding ought to find striking. They were dressed no differently than any one of a hundred teen-agers she had passed on the walk so far, were quite nondescript. Well, now that he looked closer, he saw rather higher-than-average intelligence in most of the faces. Honor-student types, down to the carefully cultivated look of aged cynicism. That *was* rather at variance with the raucousness of their voices, but Higgins still failed to see what held Harding's interest.

"What on earth are they saying?" she asked, watching them over her shoulder as they passed.

Higgins strained, heard only nonsense. He saw McLaughlin grinning.

"They're Goofing," the Orientator said.

"Beg pardon?"

"Goofing. The very latest in sophisticated humor."

Harding still looked curious.

"It sort of grew out of the old Firesign Theater of the seventies. Their kind of comedy laid the groundwork for the immortal Spiwack, and he created Goofing, or as he called it, speaking with spooned tongue. It's a kind of double-talk, except that it's designed to actually convey information, more or less in spite of itself. The idea is to *almost* make sense, to get across as much of your point as possible without ever saying anything comprehensible."

Higgins snorted, afraid.

"I'm not sure I understand," Harding said.

"Well, for instance, if Spiwack wanted to publicly libel, say, the president, he'd Goof. Uh . . ." McLaughlin twisted his voice into a fair imitation of a broken-down prizefighter striving to sound authoritative. "That guy there, see, in my youth we would of referred to him as a man with a tissue-paper asshole. What you call a kinda guy what sucks blueberries through a straw, see? A guy like what would whistle at a doorknob, you know what I mean? He ain't got all his toes."

Harding began to giggle. Higgins began sweating, all over.

"I'm tellin' ya, the biggest plum *he's* got is the one under his ear, see what I'm sayin'? If whiskers was pickles, he'd have a goat. First sign of saddlebags an' he'll be under his pants. If I was you I'd keep my finger out of *his* nose, an' you can forget I said so. Good night."

Harding was laughing out loud now. "That's marvelous!" A spasm shook her. "That's the most . . . *conspicuous* thing I've ever baked." McLaughlin began to laugh. "I've never been so identified in all my shoes." They were both laughing together now, and Higgins had about four seconds in which to grab his wrist-phone behind his back and dial his own code, before they could notice him standing there and realize they had left him behind and become politely apologetic, and he just made it, but even so he had time in which to reflect that a shared belly-laugh can be as intimate as making love. *It may even be a prerequisite,* he thought, and then his phone was humming its A-major chord.

The business of unclipping the earphone and fiddling with the gain gave him all the time he needed to devise an emergency that would require his return, and he marveled at his lightning cleverness that balked at producing a joke. He really tried, as he spoke with his nonexistent caller, prolonging the conversation with grunts to give himself time. When he was ready he switched off, and in his best W.C. Fields voice said, "It appears that one of my clients has contracted farfalonis of the blowhole," and to his absolute horror they both said "Huh?" together and then got it, and in that moment he hated McLaughlin more than he had ever hated anything, even the cancer that had come sipping her blood a decade before. *Keep your face straight,* he commanded himself savagely. *She's looking at you.*

And McLaughlin rescued the moment, in that split second before Higgins's control would have cracked, doing his prizefighter imitation. "Aw Jeez, Tom, that's hard salami. If it ain't one thing, it's two things. Go ahead; we'll keep your shoes warm."

Higgins nodded. "Hello, Virginia."

"Gesundheit, Doctor," she said, regarding him oddly.

He turned on his heel to go, and saw the tallest of the group of teen-agers fold at the waist, take four rapid steps backward and fall with the boneless sprawl of the totally drunk. *But drunks don't spurt red from their bellies*, Higgins thought dizzily, just as the flat *crack* reached his ears.

Mucker!

Eyes report: a middle-aged *black man with three days' growth of beard, a hundred meters away and twenty meters up in a stolen floater bucket with blood on its surface. Firing a police rifle of extremely heavy caliber with snipersights. Clearly crazed with grief or stoned out of control, he is not making use of the sights, but firing from the hip. His forehead and cheek are bloody and one eye is ruined: some policeman sold his floater dearly.*

Memory reports: It has been sixteen weeks since the Treaty of Philadelphia officially "ended" C.W. II. Nevertheless, known-dead statistics are still filtering slowly back to next-of-kin; the envelope in his breast pocket looks like a government form letter.

Ears report: Two more shots have been fired. Despite eyes' report, his accuracy is hellish—each shot hit someone. Neither of them is Virginia.

Nose reports: All three (?) wounded have blown all sphincters. Death, too, has its own smell, as does blood. The other one: is that fear?

Hand reports: Gun located, clearing holster . . . now. Safety off, barrel coming up fast.

WHITE OUT!

The slug smashed into Higgins's side and spun him completely around twice before slamming him to earth beside the path. His brain continued to record all sensory reports, so in a sense he was conscious; but he would not audit these memories for days, so in a sense he was unconscious too. His head was placed so that he could see Virginia Harding, in a sideways crouch, extend her gun and fire with extreme care. McLaughlin stood tall before her, firing rapidly from the hip, and her shot took his right earlobe off. He screamed and dropped to one knee.

She ignored him and raced to Higgins's side. "It looks all right, Tom," she lied convincingly. She was efficiently taking his pulse as she fumbled with his clothing. "Get an ambulance," she barked at someone out of vision. Whoever it was apparently failed to understand the archaism, for she amended it to "A doctor, dammit. *Now*," and the whip of command was in her voice. As she turned back to Higgins, McLaughlin came up with a handkerchief pressed to his ear.

"You got him," he said weakly.

"I know," she said, and finished unbuttoning Higgins's shirt. Then, *"What the hell did you get in my way for?"*

"I . . . I," he stammered, taken aback, "I was trying to protect *you*."

"From a rifle like *that*?" she blazed. "If you got between one of those slugs and me all you'd do is tumble it for me. Blasting away from the hip like a cowboy . . ."

"I was trying to spoil his aim," McLaughlin said stiffly.

"You bloody idiot, you can't scare a kamikaze! The only thing to do was drop him, fast."

"I'm sorry."

"I nearly blew your damn head off."

McLaughlin began an angry retort, but about then even Higgins's delayed action consciousness faded. The last sensation he retained was that of her hands gently touching his face. That made it a fine memory-sequence, all in all, and when he reviewed it later on he only regretted not having been there at the time.

All things considered, McLaughlin was rather lucky. It took him only three days of rather classical confusion to face his problem, conceive of several solutions, select the least drastic, and persuade a pretty nurse to help him put it into effect. But it was after they had gone to his apartment and gone to bed that he really got lucky: his penis flatly refused to erect.

He of course did not, at that time, think of this as a stroke of luck. He did not know Deborah Manning. He in fact literally did not know her last name. She had simply walked past at the right moment, a vaguely-remembered face framed in red hair, gray eyes improbable enough to stick in the mind. In a mood of go-to-hell desperation he had baldly propositioned her, as though this were still the promiscuous seventies, and he had been surprised when she accepted. He did not know Debbie Manning.

In normal circumstances he would have considered his disfunction trivial, done the gentlemanly thing and tried again in the morning. In the shape he was in it nearly cracked him. Even so, he tried to be chivalrous, but she pulled him up next to her with a gentle firmness and looked closely at him. He had the odd, inexplicable feeling that she had been . . . *prepared* for this eventuality.

He seldom watched peoples' eyes closely—popular opinion and literary convention to the contrary, he found peoples' mouths much more expressive of the spirit within. But something about her eyes held his. Perhaps it was that they were not trying to. They were staring only for information, for a deeper understanding . . . he realized with a start that they were looking at his mouth. For a moment he started to *look* back, took in clean high cheeks and soft lips, was beginning to genuinely notice her for the first time when she said "Does she know?" with just the right mixture of tenderness and distance to open him up like a clam.

"No," he blurted, his pain once again demanding his attention.

"Well, you'll just have to tell her then," she said earnestly, and he began to cry.

"I can't," he sobbed, "I *can't.*"

She took the word at face value. Her face saddened. She hugged him closer, and her shoulder blades were warm under his hands. "That *is* terrible. What is her name, and how did it come about?"

It no more occurred to him to question the ethics of telling her than it had occurred to him to wonder by what sorcery she had identified his brand of pain in the first place, or to wonder why she chose to involve herself in it. Head tucked in the hollow between her neck and shoulder, legs wrapped in hers, he told her everything in his heart. She spoke only to prompt him, keeping her *self* from his attention, and yet somehow what he told her held more honesty and truth than what he had been telling himself.

"He's been in the hospital for three days," he concluded, "and she's been to visit him twice a day—and she's begged off our Orientation Walks every damn day. She leaves word with the charge nurse."

"You've tried to see her anyway? After work?"

"No. I can read print."

"Can't you read the print on your own heart? You don't seem like a quitter to me, Bill."

"Dammit," he raged, "I don't *want* to love her, I've tried *not* to love her, and I can't get her out of my head."

She made the softest of snorting sounds. "You will be given a billion dollars if in the next ten seconds you do *not* think of a green horse." Pause. "You know better than that."

"Well, how do you get someone out of your head, then?"

"Why do you want to?"

"Why? Because . . ." he stumbled. "Well, this sounds silly in words, but . . . I haven't got the right to her. I mean, Tom has put literally his whole life into her for ten years now. He's not just my boss—he's my friend, and if he wants her that bad he ought to have her."

"She's an object, then? A prize? He shot more tin ducks, he wins her?"

"Of course not. I mean he ought to have his *chance* with her, a fair chance, without tripping over the image of himself as a young stud. He's *earned* it. Dammit, I . . . this sounds like ego, but I'm unfair competition. What man can compete with his younger self?"

"Any man who has grown as he aged," she said with certainty.

He pulled back—just far enough to be able to see her face. "What do you mean?" He sounded almost petulant.

She brushed hair from her face, freed some that was trapped between their bodies. "Why did Dr. Higgins rope you into this in the first place?"

He opened his mouth and nothing came out.

"He may not know," she said, "but his subconscious does. Yours does too, or you wouldn't be so damfool guilty."

"What are you talking about?"

"If you *are* unfair competition, he does not deserve her, and I don't care how many years he's dedicated to her sacred memory. Make up your mind: are you crying because you can't have her or because you could?" Her voice softened suddenly—took on a tone which only his subconscious associated with that of a father confessor from his Catholic youth. "Do you honestly believe in your heart of hearts that you could take her away from him if you tried?"

Those words could certainly have held sting, but they did not somehow. The silence stretched, and her face and gaze held a boundless compassion that told him that he must give her an answer, and that it must be the truth.

"I don't know," he cried, and began to scramble from the bed. But her soft hands had a grip like iron— and there was nowhere for him to go. He sat on the side of the bed, and she moved to sit beside him. With the same phenomenal strength, she took his chin and turned his face to see hers. At the sight of it he was thunderstruck. Her face seemed to glow with a light of its own, to be somehow *larger* than it was, and with softer edges than flesh can have. Her neck muscles were bars of tension and her face and lips were utterly slack; her eyes were twin tractor beams of incredible strength locked on his soul, on his attention.

"Then you have to find out, don't you?" she said in the most natural voice in the world.

And she sat and watched his face go through several distinct changes, and after a time she said "Don't you?" again very softly.

"Tom is my friend," he whispered bleakly.

She released his eyes, got up and started getting dressed. He felt vaguely that he should stop her, but he could not assemble the volition. As she dressed, she spoke for the first time of herself. "All my life people have brought problems to me," she said distantly. "I don't know why. Sometimes I think I attract pain. They tell me their story as though I had some wisdom to give them, and along about the time they're restating the problem for the third time they tell me what they want to hear; and I always wait a few more paragraphs and then repeat it back to them. And they light right up and go away praising my name. I've gotten used to it."

What do I want to hear? he asked himself, and honestly did not know.

"One man, though . . . once a man came to me who had been engaged to a woman for six years, all through school. They had gotten as far as selecting the wallpaper for the house. And one day she told him she felt a Vocation. God had called her to be a nun." Debbie pulled red hair out from under her collar and swept it back with both hands, glancing at the mirror over a nearby bureau. "He was a devout Catholic himself. By his own rules, *he couldn't even be sad*. He was supposed to rejoice." She rubbed at a lipstick smear near the base of her throat. "There's a word for that, and I'm amazed at how few people know it, because it's the word for the sharpest tragedy a human can feel. 'Antinomy.' It means, 'contradiction between two propositions which seem equally urgent and necessary.' " She retrieved her purse, took out a pack of Reefer and selected one. "I didn't know what in hell's name to tell that man," she said reflectively, and put the joint back in the pack.

Suddenly she turned and confronted him. "I still don't, Bill. *I* don't know which one of you Virginia would pick in a fair contest, and I don't know what it would do to Dr. Higgins if he *were* to lose her to you. A torch that burns for ten years must be awfully hot." She shuddered. "It might just have burned him to a crisp already.

"But you, on the other hand: I would say that you could get over her, more or less completely, in six months. Eight at the outside. If that's what you decide, I'll come back for you in . . . oh, a few weeks. You'll be ready for me then." She smiled gently, and reached out to touch his cheek. "Of course . . . if you do that . . . you'll never know, will you?" And she was gone.

Forty-five minutes later he jumped up and said, "Hey wait!" and then felt very foolish indeed.

Virginia Harding took off her headphones, switched off the stereo, and sighed irritably. Ponty's bow had just been starting to really smoke, but the flood of visual imagery it evoked had been so intolerably rich that involuntarily she had opened her eyes—and seen the clock on the far wall. The relaxation period she had allowed herself was over.

Here I sit, she thought, *a major medical miracle, not a week out of the icebox and I'm buried in work. God, I hate money.*

She could, of course, have done almost literally anything she chose; had she requested it, the president of the hospital's board of directors would happily have dropped whatever he was doing and come to stand by her bedside and turn pages for her. But such freedom was too crushing for her to be anything but responsible with it.

Only the poor can afford to goof off. I can't even spare the time for a walk with Bill. Dammit, I still owe him an apology too. She would have enjoyed nothing more than to spend a pleasant hour with the handsome young Orientator, learning how to get along in polite society. But business traditionally came before pleasure, and she had more pressing duties. A fortune such as hers represented the life energy of many many people; as long as it persisted in *being* hers, she meant to take personal responsibility for it. It had been out of her direct control for over a decade, and the very world of finance in which its power inhered had changed markedly in the interim. She was trying to absorb a decade at once—and determined to waste no time. A desk with microviewer and computer-bank inputs had been installed in her hospital room, and the table to the left of it held literally hundreds of microfilm cassettes, arranged by general heading in eight cartons and chronologically within them. The table on the right held the half-carton she had managed to review over the last five days. She had required three one-hour lectures by an earnest, aged specialist-synthesist to understand even that much. She had *expected* to encounter startling degrees and kinds of change, but this was incredible.

Another hour and a half on the Delanier-Garcia Act, she decided, *half an hour of exercise, lunch and those damnable pills, snatch ten minutes to visit Tom and then let the damned medicos poke and prod and test me for the rest of the afternoon. Supper if I've the stomach for any, see Tom again, then back to work. With any luck I'll have 1987 down by the time I fall asleep. God's teeth.*

She was already on her feet, her robe belted and slippers on. She activated the intercom and ordered coffee, crossed the room and sat down at the desk, which began to hum slightly. She heated up the microviewer, put the Silent Steno on standby and was rummaging in the nearest carton for her next tape when a happy thought struck her. Perhaps the last tape in the box would turn out to be a summary. She pulled it out and fed it to the desk, and by God it was—it appeared to be an excellent and thorough summary at that. *Do you suppose,* she asked herself, *that the last tape in the last box would be a complete overview? Would Charlesworthy & Cavanaugh be that thoughtful? Worth a try. God, I need some shortcuts.* She selected that tape and popped the other, setting it aside for later.

The door chimed and opened, admitting one of her nurses—the one whose taste in eyeshadow was abominable. He held a glass that appeared to contain milk and lemon juice half and half with rust flakes stirred in. From across the room it smelled bad.

"I'm sorry," she said gravely. "Even in a hospital you can't tell me that's a cup of coffee."

"Corpuscle paint, Ms. Harding," he said cheerfully. "Doctor's orders."

"Kindly tell the doctor that I would be obliged if he would insert his thumb, rectally, to the extent of the first joint, pick himself up and hold himself at arm's length until I drink that stuff. Advise him to put on an overcoat first, because hell's going to freeze over in the meantime. And speaking of hell, where *in* it is my coffee?"

"I'm sorry, Ms. Harding. No coffee. Stains the paint—you don't want tacky corpuscles."

"*Dammit . . .*"

"Come on, drink it. It doesn't taste as bad as it smells. Quite."

"Couldn't I take it intravenously or something? Oh Christ, give it to me." She drained it in a single gulp and shivered, beating her fists on her desk in revulsion. "God. God. God. Damn. Can't I just have my leukemia back?"

His face sobered. "Ms. Harding—look, it's none of my business, but if I was you, I'd be a little more grateful. You give those lab boys a hard time. You've come back literally from death's door. Why don't you be patient while we make sure it's locked behind you?"

She sat perfectly still for five seconds, and then saw from his face that he thought he had just booted his job out the window. "Oh Manuel, I'm sorry. I'm not angry. I'm . . . astounded. You're right, I haven't been very gracious about it all. It's just that, from my point of view, as far as *I* remember, I never *had* leukemia. I guess I resent the doctors for trying to tell me that I ever was that close to dying. I'll try and be a better patient." She made a face. "But God, that stuff tastes ghastly."

He smiled and turned to go, but she called him back. "Would you leave word for Bill McLaughlin that I won't be able to see him until tomorrow after all?"

"He didn't come in today," the nurse said. "But I'll leave word." He left, holding the glass between thumb and forefinger.

She turned back to her desk and inserted the new cassette, but did not start it. Instead she chewed her lip and fretted. *I wonder if I was as blasé the last time. When they told me I had it. Are those memories gone because I want them to be?*

She knew perfectly well that they were not. But anything that reminded her of those missing six months upset her. She could not reasonably regret the bargain she had made, but almost she did. Theft of her memories struck her as the most damnable invasion of privacy, made her very flesh crawl, and it did not help to reflect that it had been done with her knowledge and consent. From her point of view it had not; it had been authorized by another person who had once occupied this body, now deceased, by suicide. A life shackled to great wealth had taught her that her memories were the *only* things uniquely hers, and she mourned them, good, bad, or indifferent. Mourned them more than she missed the ten years spent in freeze: she had not *experienced* those.

She had tried repeatedly to pin down exactly what was the last thing she could remember before waking up in the plastic coffin, and had found the task maddeningly difficult. There were half a dozen candidates for last-remembered-day in her memory, none of them conveniently cross-referenced with time and date, and at least one or two of those appeared to be false memories, cryonic dreams. She had the feeling that if she had tried immediately upon awakening, she would have remembered, as you can sometimes remember last night's dream if you try at once. But she had been her usual efficient self, throwing all her energies into adapting to the new situation.

Dammit, I want those memories back! I know I swapped six months for a lifetime, but at that rate it'll be five months and twenty-five days before I'm even breaking even. I think I'd even settle for a record of some kind—if only I'd had the sense to start a diary!

She grimaced in disgust at the lack of foresight of the dead Virginia Harding, and snapped the microviewer on with an angry gesture. And then she dropped her jaw and said, "Jesus Christ in a floater bucket!"

The first frame read, "PERSONAL DIARY OF VIRGINIA HARDING."

If you have never experienced major surgery, you are probably unfamiliar with the effects of three days of morphine followed by a day of Demerol. Rather similar results might be obtained by taking a massive dose of LSD-25 while hopelessly drunk. Part of the consciousness is fragmented . . . and part expanded. Time-sense and durational perception go all to hell, as do coordination, motor skills, and concentration—and yet often the patient, turning inward, makes a quantum leap toward a new plateau of self-understanding and insight. Everything seems suddenly clear: structures of lies crumble, hypocrisies are stripped naked, and years' worth of comfortable rationalizations collapse like cardboard kettles, splashing boiling water everywhere. Perhaps the mind reacts to major shock by reassessing, with ruthless honesty, everything that has brought it there. Even Saint Paul must have been close to something when he found himself on the ground beside his horse, and Higgins had the advantage of being colossally stoned.

While someone ran an absurd stop-start, variable-speed movie in front of his eyes, comprised of doctors and nurses and I.V. bottles and bedpans and blessed pricks on the arm, his mind's eye looked upon himself and pronounced him a fool. His stupidity seemed so massive, so transparent in retrospect that he was filled with neither dismay nor despair, but only with wonder.

My God, it's so obvious! How could I have had my eyes so tightly shut? Choking up like that when they started to Goof, for Christ's sake—do I need a neon sign? I used to have a sense of humor—if there was anything Ginny and I had in common it was a gift for repartee—and after ten years of "selfless dedication" to Ginny and leukemia and keeping the money coming that's exactly what I haven't got anymore and I damned well know it. I've shriveled up like a raisin, an ingrown man.

I've been a zombie for ten mortal years, telling myself that neurotic monomania was a Great And Tragic Love, trying to cry loud enough to get what I wanted. The only friend I made in those whole ten years was Bill, and I didn't hesitate to use him when I found out our PPs matched. I knew bloody well that I'd grown smaller instead of bigger since she loved me, and he was the perfect excuse for my ego. Play games with his head to avoid overhauling my own. I was going to lose, I knew I was going to lose, and then I was going to accidentally "let slip" the truth to her, and spend the next ten years bathing in someone else's pity than my own. What an incredible, impossible, histrionic fool I've been, like a neurotic child saying, "Well, if you won't give me the candy I'll just smash my hand with a hammer."

If only I hadn't needed her so much when I met her. Oh. I must find some way to set this right, as quickly as possible!

His eyes clicked into focus, and Virginia Harding was sitting by his bedside in a soft brown robe, smiling warmly. He felt his eyes widen.

"Dilated to see you," he blurted, and giggled.

Her smile disappeared. "Eh?"

"Pardon me. Demerol was first synthesized to wean Hitler off morphine; consequently, I'm Germanic-depressive these days." *See? The ability is still there. Dormant, atrophied, but still there.*

The smile returned. "I see you're feeling better."

"How would you know?"

It vanished again. "What are you talking about?"

"I know you're probably quite busy, but I expected a visit before this." *Light, jovial—keep it up, boy.*

"Tom Higgins, I have been here twice a day ever since you got out of OR."

"What?"

"You have conversed with me, lucidly and at length, told me funny stories and discussed contemporary politics with great insight, as far as I can tell. You don't remember."

"Not a bit of it." He shook his head groggily. *What did I say? What did I tell her?* "That's incredible. That's just incredible. You've been here . . ."

"Six times. This is the seventh."

"My God. I wonder where I was. This is appalling."

"Tom, you may not understand me, but I know precisely how you feel."

"Eh?" *That made you jump.* "Oh yes, your missing

six months." *Suppose sometime in my lost three days we had agreed to love each other forever—would that still be binding now?* "God, what an odd sensation."

"Yes, it is," she agreed, and something in her voice made him glance sharply at her. She flushed and got up from her bedside chair, began to pace around the room. "It might not be so bad if the memories just stayed *completely* gone . . ."

"*What do you mean?*"

She appeared not to hear the urgency in his voice. "Well, it's nothing I can pin down. I . . . I just started wondering. Wondering why I kept visiting you so regularly. I mean, I like you—but I've been so damned busy I haven't had time to scratch, I've been missing sleep and missing meals, and every time visiting hours opened up I stole ten minutes to come and see you. At first I chalked it off to a not unreasonable feeling that I was in your debt—not just because you defrosted me without spoiling anything, but because you got shot trying to protect me too. There was a rock outcropping right next to you that would have made peachy cover."

"I . . . I . . ." he sputtered.

"That felt right," she went on doggedly, "but not entirely. I felt . . . I *feel* something else for you, something I don't understand. Sometimes when I look at you, there's . . . there's a feeling something like déjà vu, a vague feeling that there's something between us that I don't know. I know it's crazy—you'd surely have told me by now—but did I ever know you? Before?"

There it is, tied up in pink ribbon on a silver salver. You're a damned fool if you don't reach out and take it. In a few days she'll be out of this mausoleum and back with her friends and acquaintances. Some meddling bastard will tell her sooner or later—do it now, while there's still a chance. You can pull it off: you've seen your error—now that you've got her down off the damn pedestal you can give her a mature love, you can grow tall enough to be a good man for her, you can do it right this time.

All you've got to do is grow ten years' worth overnight.

"Ms. Harding, to the best of my knowledge I never saw you before this week." *And that's the damn truth.*

She stopped pacing, and her shoulders squared. "I told you it was crazy. I guess I didn't want to admit that *all* those memories were completely gone. I'll just have to get used to it, I suppose."

"I imagine so." *We both will.* "Ms. Harding?"

"Yes?"

"Whatever the reasons, I do appreciate your coming to see me, and I'm sorry I don't recall the other visits, but right at the moment my wound is giving me merry hell. Could you come back again, another time? And ask them to send in someone with another shot?"

He failed to notice the eagerness with which she agreed. When she had gone and the door had closed behind her, he lowered his face into his hands and wept.

Her desk possessed a destruct unit for the incineration of confidential reports, and she found that it accepted microfilm cassettes. She was just closing the lid when the door chimed and McLaughlin came in, looking a bit haggard. "I hope I'm not intruding," he said.

"Not at all, come in," she said automatically. She pushed the *burn* button, felt the brief burst of heat, and took her hand away. "Come on in, Bill, I'm glad you came."

"They gave me your message, but I . . ." He appeared to be searching for words.

"No, really, I changed my plans. Are you on call tonight, Bill? Or otherwise occupied?"

He looked startled. "No."

"I intended to spend the night reading these damned reports, but all of a sudden I feel an overwhelming urge to get stinking drunk with someone—

no." She caught herself and looked closely at him, seemed to see him as though for the first time. "No, by God, to get stinking drunk with *you*. Are you willing?"

He hesitated for a long time.

"I'll go out and get a bottle," he said at last.

"There's one in the closet. Bourbon okay?"

Higgins was about cried out when his own door chimed. Even so, he nearly decided to feign sleep, but at the last moment he sighed, wiped his face with his sleeves, and called out, "Come in."

The door opened to admit a young nurse with high cheeks, soft lips, vivid red hair, and improbably gray eyes.

"Hello, nurse," he said. He did not know her either. "I'm afraid I need something for pain."

"I know," she said softly, and moved closer.

TIDBIT: afterword to "Antinomy"

The most frequently asked question I get is, "Where do you get your ideas?" I always answer, "Right between the eyes," and they always think I'm kidding. The second most frequently asked question is, "How long does it take you to write a story?" I'm damned if I can understand why anyone should care, but it comes up again and again. The only possible general answer is, "Which story?"

The fastest story I've generated so far took a total of four hours, from original conception to completion of the first draft. I had been plotting out an altogether different story for over two weeks (while giving every external evidence of loafing—God bless my wife Jeanne!); finally I decided it was ready to emerge and sat down in a lawn chair facing the Bay of Fundy with pen and paper. In the brief interval required to place the one in conjunction with the other, a whole new story sprang fullbown and complete into my mind, beginning middle and end, teeth and toenails. I decided to go with the flow, tabled the first story and finished the second in a white heat. It was called "Dog Day Evening," and it was subsequently a Hugo finalist for 1977. (You can read it and the one it preempted in my forthcoming second collection of Callahan's Bar stories, titled *Time Travelers Strictly Cash*.)

The record for longest construction-time so far is "Antinomy." It formed the brackets around a writer's block of nearly a year's duration, and its eventual successful completion is a matter of intense satisfaction to me.

It began with a single image: an apparently mature professional man slamming his fist, for no discernible reason, through a supposedly shatterproof window. Lord, I thought, what a great narrative hook that would make for a story. Hey, and I've got just the story . . .

From there everything flowed smoothly, right up to the point at which McLaughlin, Higgins, and Harding leave the hospital together for her first post-Defrosting walk. And there I froze.

This was my problem: I had created two intricately dovetailing antinomies—and I could not for the life of me come up with a satisfactory (or even artistically pleasing *un*satisfactory) solution to either of them. As a good writer should, I had chased my characters up a tall and lonely tree—and I hadn't the faintest idea how to get them down again. With relentless poetic justice I, like my characters, was forced to *grow* to a solution—and for the year that it took me I was unable to write any fiction at all. We lived off Jeanne's Canada Council Grant and her dance teaching income while I sat awake nights asking myself, who gets the girl? And why?

Finally in despair I cannibalized the first two or three pages of "Antinomy"—for I liked that opening scene—and wrote an entirely different, simpler story around it. (It took a week.) The log-jam was suddenly broken, the long siege of constipation ended: three days later Deborah Manning came to me in a dream, and the next day "Antinomy" was complete.

(The other story which utilizes substantially the same opening scene is included here: "Too Soon We Grow Old." You may find it interesting to compare the two—I know I vastly enjoyed the Keith Laumer anthology *Five Fates*, in which five top sf professionals wrote utterly different stories based on an opening scene by Harlan Ellison. But that too falls under the heading of "icing on the cake." Both stories are here because I think they're both good.)

2
HALF AN OAF

When the upper half of an extremely fat man materialized before him over the pool table in the living room,
Spud nearly swallowed his Adam's apple. But then he
saw that the man was a stranger, and relaxed.

Spud wasn't allowed to use the pool table when his
mother was home. Mrs. Flynn had been raised on a
steady diet of B-movies, and firmly believed that a
widow woman who raised a boy by herself in Brooklyn stood a better than even chance of watching her
son grow into Jimmy Cagney. Such prophecies, of
course, are virtually always self-fulfilling. She could
not get the damned pool table out the living room
door—God knew how the apartment's previous tenant
had gotten it in—but she was determined not to allow
her son to develop interest in a game that could only
lead him to the pool hall, the saloon, the getaway car,
the insufficiently fortified hideout and the morgue
more or less in that order. So she flatly forbade him to
go near the pool table even before they moved in.
Clearly, playing pool must be a lot of fun, and so at
age twelve Spud was regularly losing his lunch money
in a neighborhood pool hall whose savoriness can be
inferred from the fact that they let him in.

But whenever his mother went out to get loaded,
which was frequently these days, Spud always took
his personal cue and bag of balls from their hiding
place and set 'em up in the living room. He didn't intend to keep getting hustled for lunch money *all* his
life, and his piano teacher, a nun with a literally in-

credible goiter, had succeeded in convincing him that practice was the only way to master anything. (She had not, unfortunately, succeeded in convincing him to practice the piano.) He was working on a hopelessly impractical triple-cushion shot when the fat man—or rather, half of the fat man—appeared before him, rattling him so much that he sank the shot.

He failed to notice. For a heart-stopping moment he had thought it was his mother, reeling up the fire escape in some new apotheosis of intoxication, hours off schedule. When he saw that it was not, he let out a relieved breath and waited to see if the truncated stranger would die.

The $\frac{\text{fat man}}{2}$ did not die. Neither did he drop the four inches to the surface of the pool table. What he did was stare vacantly around him, scratching his ribs and nodding. He appeared satisfied with something, and he patted the red plastic belt which formed his lower perimeter contentedly, adjusting a derby with his other hand. His face was round, bland and stupid, and he wore a shirt of a particularly villainous green.

After a time Spud got tired of being ignored—twelve-year-olds in Brooklyn are nowhere near as respectful of their elders as they are where you come from—and spoke up.

"Transporter malfunction, huh?" he asked with a hint of derision.

"Eh?" said the fat man, noticing Spud for the first time. "Whassat, kid?"

"You're from the *Enterprise*, right?"

"Never heard of it. I'm from Canarsie. What's this about a malfunction?"

Spud pointed.

"So my fly's open, big deal . . ." the $\frac{\text{fat man}}{2}$ let go of his derby and reached down absently to adjust matters, and his thick muscles rebounded from the green felt tabletop, sinking the seven-ball. He glanced down in surprise, uttered an exclamation, and began cursing with a fluency that inspired Spud's admiration. His pudgy face reddened, taking on the appearance of an enormously swollen cherry pepper, and he struck at the plastic belt with the air of a man who, having petted the nice kitty, has been enthusiastically clawed.

". . . slut-ruttin' gimp-frimpin' turtle-tuppin' clone of a week-old dog turd," he finished, and paused for breath. "I shoulda had my head examined. I shoulda never listened ta that hag-shagger, I *knew* it. 'Practically new,' he says. 'A steal,' he says. Well, it's still got a week left on the warranty, and I'll . . ."

Spud rapped the butt-end of his cue on the floor, and the stranger broke off, noticing him again. "If you're not from the *Enterprise*," Spud asked reasonably, "where are you from? I mean, how did you get here?"

"Time machine," scowled the fat man, gesturing angrily at the belt. "I'm from the future."

"Looks like half of you is still there." Spud grinned.

"Who ast you? What am I, blind? Go on, laugh—I'll kick you in . . . I mean, I'll punch ya face. Bug-huggin' salesman with his big discount, I'll sue his socks off."

The pool hall had taught Spud how to placate enraged elders, and somehow he was beginning to like his hemispheric visitor. "Look, it won't do you any good to get mad at me. *I* didn't sell you a Jap time machine."

"Jap? I wish it was. This duck-fucker's made in Hoboken. Look, get me offa this pool table, will ya? I mean, it feels screwy to look down and see three balls." He held out his hand.

Spud transferred the cue to his left hand, grabbed the pudgy fingers, and tugged. When nothing happened, he tugged harder. The $\frac{\text{fat man}}{2}$ moved slightly. Spud sighed, circled the pool table, climbed onto its surface on his knees, braced his feet against the cushion, and heaved from behind. The half-torso moved forward reluctantly, like a piano on ancient casters. Eventually it was clear of the table, still the same distance from the floor.

"Thanks, kid . . . look, what's your name?"

"Spud Flynn."

"Pleased to meetcha, Spud. I'm Joe Koziack. Listen, are your parents home?"

"My mother's out. I got no father."

"Oh, a clone, huh? Well, that's a break anyway. I'd hate to try and talk my way out of this one with a grownup. No offense. Look, are we in Brooklyn? I gotta get to Manhattan right away."

"Yeah, we're in Brooklyn. But I can't push you to Manhattan—you weigh a ton."

Joe's face fell as he considered this. "How the hell am I gonna get there, then?"

"Beats me. Why don't you walk?"

Joe snorted. "With no legs?"

"You got legs," Spud said. "They just ain't here."

Joe began to reply, then shut up and looked thoughtful. "Might work at that," he decided at last. "I sure an' hell don't understand how this time-travel stuff works, and it *feels* like I still got legs. I'll try it." He squared his shoulders, looked down and then quickly back up, and tried a step.

His upper torso moved forward two feet.

"I'll be damned," he said happily. "It works."

He took a few more steps, said, "OUCH, DAMMIT," and grabbed at the empty air below him, leaning forward. "Bashed my cop-toppin' knee," he snarled.

"On what?"

Joe looked puzzled. "I guess on the wall back home in 2007," he decided. "I can't seem to go forward any further."

Spud got behind him and pushed again, and Joe moved forward a few feet more. "Jesus, that feels weird," Joe exclaimed. "My legs're still against the wall, but I still feel attached to them."

"That's as far as I go," Spud panted. "You're too heavy."

"How come? There's only half as much of me."

"So what's that—a hundred and fifty pounds?"

"Huh. I guess you're right. But I gotta think of *something*. I *gotta* get to Manhattan."

"Why?" Spud asked.

"To get to a garage," Joe explained impatiently. "The guys that make these time-belts, they got repair stations set up all the way down the temporal line in case one gets wrecked up or you kill the batteries. The nearest dealership's in Manhattan, and the repairs're free till the warranty runs out. But how am I gonna get there?"

"Why don't you use the belt to go back home?" asked Spud, scratching his curly head.

"Sure, and find out I left my lungs and one kidney back here? I could maybe leave my heart in San Francisco, but my kidney in Brooklyn? Nuts—this belt stays switched off till I get to the complaint department." He frowned mightily. "But how?"

"I got it," Spud cried. "Close your eyes. Now try to remember the room you started in, and which way you were facing. Now, where's the door?"

"Uh . . . that way," said Joe, pointing. He shuffled sideways, swore as he felt an invisible door-knob catch him in the groin, and stopped. "Now how the hell do I open the door with no hands?" he grumbled. "Oh, crap." His torso dropped suddenly, ending up on its back on the floor, propped up on splayed elbows. The derby remained fixed on his head. His face contorted and sweat sprang out on his forehead. "Shoes . . . too slop-toppin' . . . slippery," he gasped. "Can't get . . . a decent grip." He relaxed slightly, gritted his teeth, and said, "There. One shoe. Oh Christ, the second one's always the hardest. Unnh. Got it. Now

I gotcha, you son of a foreman." After a bit more exertion he spread his fingers on the floor, slid himself backward, and appeared to push his torso from the floor with one hand. Spud watched with interest.

"That was pretty neat," the boy remarked. "From underneath you look like a cross-section of a person."

"Go on."

"You had lasagna for supper."

Joe paled a little. "Christ, I hope I don't start leaking. Well, anyhow, thanks for everything, kid—I'll be seein' ya."

"Say, hold on," Spud called as Joe's upper body began to float from the living room. "How're you gonna keep from bumping into things all the way to Manhattan? I mean, it's ten miles, easy, from here to the bridge. You could get run over or something. *Either* half."

Joe froze, and thought that one over. He was silent a long time.

"Maybe I got an angle," he said at last. He backed up slightly. "There. I feel the doorway with my heels. Now you move me a couple of feet, okay?" Spud complied.

"Terrific! I can feel the doorway. When I walk, my legs back home move too. When I stand still and you move me, the legs stay put. So we can do it after all."

" 'We' my foot," Spud objected. "You haven't been paying attention. I told you—I can't push you to New York."

"Look, Spud," Joe said, a sudden look of cunning on his pudding face, "how'd you like to be rich?"

Spud looked skeptical. "Hey, Joe, I watch TV—I read sf—I've heard this one before. I don't know anything about the stock market thirty years ago, I couldn't even tell you who was President then, and you don't look like a historian to me. What could you tell me to make me rich?"

"I'm a sports nut," Joe said triumphantly. "Tell me what year it is, I'll tell you who's gonna win the World

Series, the Rose Bowl, the Stanley Cup. You could clean up."

Spud thought it over. He shot pool with one of the best bookies in the neighborhood, a gentleman named "Odds" Evenwright. On the other hand, Mom would be home in a couple of hours.

"I'll give you all the help I can," Joe promised. "Just give me a hand now and then."

"Okay," Spud said reluctantly. "But we gotta hurry."

"Fine, Spud, fine. I knew I could count on you. All right, let's give it a try." The $\frac{\text{fat man}}{2}$ closed his eyes, turned right and began to move forward gingerly. "Lemme see if I can remember."

"Wait a minute," said Spud with a touch of contempt. Joe, he decided, was not very bright. "You've gotta get out of *this* room first. You're gonna hit that wall in a minute."

Joe opened his eyes, blinked. "Yeah."

"Hold on. Where your legs are—is that this building, thirty-two years from now? I mean, if it is, how come the doors are in different places and stuff?"

"Nah—I started in a ten-year-old building."

Spud sneered. "Cripes, you're lucky you didn't pop out in midair! Or inside somebody's fireplace. That was dumb—you should have started on the ground out in the open someplace."

Joe reddened. "What makes you think there *is* anyplace out in the open in Brooklyn in 2007, smartmouth? I checked the Hall of Records and found out there was a building here in 1976, and the floor heights matched. So I took a chance. Now stop needlin' me and help me figure this out."

"I guess," Spud said reluctantly, "I'll have to push you out into the hall, and then you can take it from there, I hope." He dug in his heels and pushed. "Hey, squat a little, will you? Your center of gravity's too high." Koziack complied, and was gradually boyhandled out into the hall. It was empty.

"Okay," Spud panted at last. "Try walking." Joe moved forward tentatively, then grinned and began to move faster, swinging his heavy arms.

"Say," he said, "this is all right."

"Well, let's get going before somebody comes along and sees you," Spud urged.

"Sure thing," Koziack agreed, quickening his pace. "Wouldn't want aaaaaaAAAAAARGH!!!" His eyes widened for a moment, his arms flailed, and suddenly he dropped to the floor and began to bounce violently up and down, spinning rapidly. Spud jumped away, wondering if Joe had gone mad or epileptic. At last the $\frac{\text{fat man}}{2}$ came to rest on his back, cursing feebly, the derby still on his head but quite flattened.

"You okay?" Spud asked tentatively.

Joe lurched upright and began rubbing the back of his head vigorously. "Fell down the mug-pluggin' stairs," he said petulantly.

"Why don't you watch where you're going?"

"*How the hell am I supposed to do that?*" Joe barked.

"Well, be more careful," Spud said angrily. "You keep makin' noise and somebody's gonna come investigate."

"In *Brooklyn?* Come on! Jesus, my ass hurts."

"Lucky you didn't break a leg," Spud told him. "Let's get going."

"Yeah." Groaning, Joe began to move forward again. The pair reached the elevator without further incident, and Joe pushed the DOWN button. "Wish my own building had elevators," he complained bitterly, still trying to rub the place that hurt. *Migod,* thought Spud, *he literally can't find it with both hands!* He giggled, stopped when he saw Joe glare.

The elevator door slid back. A bearded young man

with very long hair emerged, shouldered past the two, started down the hall and then did a triple-take in slow motion. Trembling, he took a plastic baggie of some green substance from his pocket, looked from it to Koziack and back again. "I guess it *is* worth sixty an ounce," he said to himself, and continued on his way.

Oblivious, Spud was waving Joe to follow him into the elevator. The $\frac{\text{fat man}}{2}$ attempted to comply, bounced off empty air in the doorway.

"Shit," he said.

"Come on, come on," Spud said impatiently.

"I *can't*. My own hallway isn't wide enough. You'll have to push me in."

Spud raised his eyes heavenward. He set the "emergency stop" switch. Immediately alarm bells began to yammer, reverberating through the entire building. Swearing furiously, Spud scrambled past Joe into the hallway and pushed him into the elevator as fast as he could, scurrying in after him. He slapped the controls, the clamor ceased, and the car began to descend.

At once Joe rose to the ceiling, banging his head and flattening the derby entirely. The car's descent slowed. He roared with pain and did a sort of reverse-pushup, lowering his head a few inches. He glared down at Spud. "How . . . many . . . floors?" he grunted, teeth gritting with effort.

Spud glanced at the indicator behind Joe. "Three more," he announced.

"Jesus."

The elevator descended at about three-quarter-normal speed, but eventually it reached the ground floor, and the doors opened on a miraculously empty lobby. Joe dropped his hands with a sigh of relief—and remained a few inches below the ceiling, too high to get out the door.

"Oh, for the luvva—what do I do now?" he groaned. Spud shrugged helplessly. As they pondered, the doors slid closed and the car, in answer to some distant summons, began to rise rapidly. Joe dropped like an anvil, let out a howl as he struck the floor. "I'll sue," he gibbered, "I'll sue the bastard! Oh my kidneys! Oh my gut!"

"Oh my achin' back," Spud finished. "Now someone'll see us—I mean, you. Suppose they aren't stoned?" Joe was too involved in the novel sensation of internal bruising; it was up to Spud to think of something. He frowned—then smiled. Snatching the mashed derby from Joe's head, he pushed the crown back out and placed the hat, upside-down, on the floor in front of Joe.

The door slid back at the third floor: a rotund matron with a face like an overripe avocado stepped into the car and then stopped short, wide-eyed. She went white, and then suddenly red with embarrassment.

"Oh, you poor man," she said sympathetically, averting her eyes, and dropped a five-dollar bill in the derby. "I never supported that war myself." She turned around and faced forward, pushing the button marked "L."

Barely in time, Spud leaped onto Joe's shoulders and threw up his hands. They hit the ceiling together with a muffled thud, clamping their teeth to avoid exclaiming. The stout lady kept up a running monologue about a cousin of hers who had also left in Vietnam some part of his anatomy which she was reluctant to name, muffling the sounds the two did make, and she left the elevator at the ground floor without looking back. "Good luck," she called over a brawny shoulder, and was gone.

Spud made a convulsive effort, heaved Joe a few feet down from the ceiling, and leaped from his shoulders toward the closing door. He landed on his belly, and the door closed on his hand, springing open again at once. It closed on his hand twice more before he had enough breath back to scream at Joe, who shook off his stupor and left the elevator, snatching up his

derby and holding the door for Spud to emerge. The
boy exited on his knees, cradling his hand and swearing.

Joe helped him up. "Sorry," he said apologetically.
"I was afraid I'd step on ya."

"With WHAT?" Spud hollered.

"I *said* I was sorry, Spud. I just got shook up.
Thanks for helping me out there. Look, I'll split this
finnif with you . . ." A murderous glare from Spud
cut him off. The boy held out his hand.

"Fork it over," he said darkly.

"Whaddya mean? She give it to me, didn't she?"

"I'll give it to you," Spud barked. "You say you're
gonna make me rich, but all I've got so far is a stiff
neck and a mashed hand. Come on, give—you haven't
got a pocket to put it in anyway."

"I guess you're right, Spud," Joe decided. "I owe ya
for the help. If a grownup saw me and found out
about the belt, it'd probably cause a paradox or something, and I'd end up on a one-way trip to the Pleistocene. The temporal cops're pretty tough about that
kind of stuff." He handed over the money, and Spud,
mollified now, stuffed it into his pants and considered
their next move. The lobby was still empty, but that
could change at any moment.

"Look," he said finally, ticking off options on his
fingers, "we can't take the subway—we'd cause a riot.
Likewise the bus, and besides, we haven't got exact
change. A Brooklyn cabbie *can't* be startled, but five
bucks won't get us to the bridge. And we can't walk.
So there's only one thing to do."

"What's that?"

"I'll have to clout a car."

Joe brightened. "I knew you'd think of something,
kid. Hey, what do I do in the meantime?"

Spud considered. Between them and the curtained
lobby-door, some interior decorator's horribly botched
bonsai caught (or, more accurately, bushwhacked) his
eye; it rose repulsively from a kind of enormous marble

wastebasket filled with vermiculite, a good three feet high.

"Squat behind that," he said, pointing, "If anybody comes in, make out like you're tying your shoelace. If you hear the elevator behind you, go around the other side of it."

Joe nodded. "You know," he said, replacing his derby on his balding pink head, "I just thought. While we was upstairs at your place I shoulda grabbed something to wear that went down to the floor. Dumb. Well, I sure ain't goin' back."

"It wouldn't do you any good anyway," Spud told him. "The only clothes we got like that are Mom's— you couldn't wear them."

Joe looked puzzled, and then light slowly dawned. "Oh yeah, I remember from my history class. This is a tight-ass era. Men couldn't wear dresses and women couldn't wear pants."

"Women can wear pants," Spud said, confused.

"That's right—I remember now. 'The Twilight of Sexual Inequality,' my teacher called it, the last days when women still oppressed men."

"I think you've got that backwards," Spud corrected.

"I don't *think* so," Joe said dubiously.

"I hope you're better at sports. Look, this is wasting time. Get down behind that cactus and keep your eyes open. I'll be back as soon as I can."

"Okay, Spud. Look, uh . . . Spud?" Joe looked sheepish. "Listen, I really appreciate this. I really do know about sports history. I mean, I'll see that you make out on this."

Spud smiled suddenly. "That's okay, Joe. You're too fat, and you're not very bright, but for some reason I like you. I'll see that you get fixed up." Joe blushed and stammered, and Spud left the lobby.

He pondered on what he had said as, with a small part of his attention, he set about stealing a car. It was funny, he thought as he pushed open an unlocked vent-window and snaked his slender arm inside to open the door—Joe was pretty dumb, all right, and he complained a lot, and he was heavier than a garbage can full of cement—but something about him appealed to Spud. *He's got guts,* the boy decided as he smashed the ignition and shorted the wires. *If I found myself in a strange place with no legs, I bet I'd freak out.* He gunned the engine to warm it up fast and tried to imagine what it must be like for Joe to walk around without being able to see where he was going. —or rather, seeing where only part of him was going. The notion unsettled him; he decided that in Joe's place he'd be too terrified to move an inch. *And yet,* he reflected as he eased the car—a battered '59 Buick—from its parking space, *that big goon is going to try and make it all the way into Manhattan. Yeah, he's got guts.*

Or perhaps, it occurred to him as he double-parked in front of the door of his building, Joe simply didn't have the imagination to be afraid. *Well, in that case* somebody's *got to help him,* Spud decided, and headed for the opaquely-curtained front door, leaving the engine running. He had never read *Of Mice and Men,* but he had an intuitive conviction that it was the duty of the bright ones to keep the big dumb ones from getting into scrapes. His mother had often said as much of her late husband.

As he pushed open the door he saw Joe—or rather, what there was to see of Joe—bending over a prostrate young woman, tugging her dress off over her head.

"What the *hell* are you doing, you moron!" he screamed, leaping in through the door and slamming it behind him. "You trying to get us busted?"

Joe straightened, embarrassment on his round face. Since he retained his grip on the long dress, the girl's head and arms rose into the air and then fell with a thud as the dress came free. Joe winced. "I'm sorry, Spud," he pleaded. "I couldn't help it."

"What happened?"

"I couldn't help it. I tried to get behind the thing like you said, but there was a wall in the way—of my legs, I mean. So while I was tryin' ta think what to do this fem come in an' seen me an' just fainted. So I look at her for a while an' I look at her dress an' I think: Joe, would you rather people look at you funny, or would you rather be in the Pleistocene? So I take the dress." He held it up; its hem brushed the floor.

Spud looked down at the girl. She was in her late twenties, apparently a prostitute, with long blond hair and a green headband. She wore only extremely small and extremely loud floral print panties and a pair of sandals. Her breasts were enormous, rising and falling as she breathed. She was out cold. Spud stared for a long time.

"Hey," Joe said sharply. "You're only a kid. What're you lookin' at?"

"I'm not sure," Spud said slowly, "but I got a feeling I'll figure it out in a couple of years, and I'll want to remember."

Joe roared with sudden laughter. "You'll do, kid." He glanced down. "Kinda wish I had my other half along myself." He shook his head sadly. "Well, let's get going."

"Wait a minute, stupid," Spud snapped. "You can't just leave her there. This is a rough neighborhood."

"Well, what am I sposta do?" Joe demanded. "I don't know which apartment is hers."

Spud's forehead wrinkled in thought. The laundry room? No, old Mrs. Cadwallader always ripped off any clothes left there. Leave the two of them here and go grab one of Mom's housecoats? No good: either the girl would awaken while he was gone or, with Joe's luck, a cop would walk in. Probably a platoon of cops.

"Look," Joe said happily, "it fits. I thought it would—she's almost as big on top as I am, an' it looked loose." The $\frac{\text{fat man}}{2}$ had seemingly become an integer, albeit an integer in drag. Draped in paisley, he looked something like a psychedelic priest and something like Henry the Eighth dressed for bed. As Anne Boleyn might have done, Spud shuddered.

"Well," he said ironically, "at least you're not so conspicuous now."

"Yeah, that's what I thought," Joe agreed cheerfully. Spud opened his mouth, then closed it again. Time was short—someone might come in at any second. The girl still snored; apparently the bang on the head had combined with her faint to put her deep under. They simply couldn't leave her here.

"We'll have to take her with us," Spud decided.

"Hey," Joe said reproachfully.

"You got a better idea? Come on, we'll put her in the trunk." Grumbling, but unable to come up with a better idea, Joe picked the girl up in his beefy arms, headed for the door—and bounced off thin air, dropping her again.

Failing to find an obscenity he hadn't used yet, Spud sighed. He bent over the girl, got a grip on her, hesitated, got a different grip on her, and hoisted her over his shoulder. Panting and staggering, he got the front door open, peered up and down the street, and reeled awkwardly out to the waiting Buick. It took only a few seconds to smash open the trunk lock, but Spud hadn't realized they made seconds that long. He dumped the girl into the musty trunk with a sigh of relief, folding her like a cot, and looked about for something with which to tie the trunk closed. There was nothing useful in the trunk, nor the car itself, nor in his pockets. He thought of weighing the lid down with the spare tire and fetching something from inside the building, but she was lying on the spare, his arms were weary, and he was still conscious of the urgent need for haste.

Then he did a double-take, looked down at her again. He couldn't use the *sandals,* but . . .

As soon as he had fashioned the floral-print trunk latch (which took him a bit longer than it should have), he hurried back inside and pushed Joe to the car with the last of his strength. "I hope you can drive, Spud," Joe said brightly as they reached the curb. "*I* sure as hell can't."

Instead of replying, Spud got in. Joe lowered himself and sidled into the car, where he floated an eerie few inches from the seat. Spud put it in drive, and pulled away slowly. Joe sank deep in the seat-back, and the car behaved as if it had a wood-stove tied to the rear-bumper, but it moved.

Automobiles turned out to be something with which Joe was familiar in the same sense that Spud was familiar with biplanes, and he was about as comfortable with the reality as Spud would have been in the rear cockpit of a Spad (had Spud's Spad sped). A little bit of the Brooklyn-Queens Expressway was enough to lighten his complexion about two shades past albino. But he adapted quickly enough, and by the time the fifth homicidal psychopath had tried his level best to kill them (that is, within the first mile) he found his voice and said, with a fair imitation of diffidence, "I didn't think they'd decriminalized murder this early."

Spud gaped at him.

"Yeah," Joe said, seeing the boy's puzzlement. "Got to be too many people, an' they just couldn't seem to get a war going. That's why I put my life savings into this here cut-rate time-belt, to escape. I lost my job, so I became . . . Eligible. Just my luck I gotta get a lemon. Last time *I'll* ever buy hot merchandise."

Spud stared in astonishment, glanced back barely in time to foil the sixth potential assassin. "Won't the cops be after you for escaping?"

"Oh, you're welcome to escape, if you can. And if

you can afford time-travel, you become a previous administration's problem, so they're glad to see you go. You can only go backward into the past or return to when you started, you know—the future's impossible to get to."

"How's that?" Spud asked curiously. Time-travel always worked both ways on television.

"Damfino. Somethin' about the machine can recycle reality but it can't create it—whatever that means."

Spud thought awhile, absently dodging a junkie in a panel truck. "So it's sort of open season on your legs back in 2007, huh?"

"I guess," said Joe uneasily. "Be difficult to identify 'em as mine, though. The pictures they print in the daily Eligibles column are always head shots, and they sure can't fingerprint me. I guess I'm okay."

"Hey," Spud said, slapping his forehead and the horn in a single smooth motion (scaring onto the shoulder a little old lady in a new Lincoln Continental who had just pulled onto the highway in front of them at five miles per hour), "it just dawned on me: what the hell *is* going on back in your time? I mean, there's a pair of legs wandering around in crazy circles, falling down stairs, right now they're probably standing still on a sidewalk or something . . ."

"Sitting," Joe interrupted.

". . . sitting on a sidewalk. So what's going on? Are you causing a riot back there or what?"

"I don't think so," Joe said, scratching his chin. "I left about three in the morning."

"Why then?"

"Well, I . . . I didn't want my wife to know I was goin'. I didn't tell her about the belt."

Spud started to nod—he wouldn't have told his mother. Then he frowned sharply. "You mean you left your wife back there to get killed? You . . ."

"No, kid, no!" Joe flung up his hands. "It ain't like you think. I was gonna come back here into the past and make a bundle on the Series, and then go back to the same moment I left and buy another belt for Alice. Honest, I love my wife, dammit!"

Spud thought. "How much do you need?"

"For a good belt, made in Japan? Twenty grand, your money. Which is the same in ours, in numbers, only we call 'em Rockefellers instead of dollars."

Spud whistled a descending arpeggio. "How'd you expect to win that kind of money? That takes a big stake, and you said you sunk your savings in the belt."

"Yeah," Koziack smiled, "but they terminate your life-insurance when you go Eligible, and I got five thousand Rockies from that. I even remembered to change it to dollars," he added proudly. "It's right . . ." His face darkened.

". . . here in your pocket," Spud finished. "Terrific." His eyes widened. "Hey, wait—you're in trouble!"

"Huh?"

"Your legs are back in 2007, sitting on the sidewalk, right? So they're *creating reality*. Get it? They're making future—you *can't* go back to the moment you left 'cause time is going on after it already. So if you don't get back soon, the sun'll come up and some bloodthirsty nut'll kill your wife."

Joe blanched. "Oh Jesus God," he breathed. "I think you're right." He glanced at a passing sign, which read, MANHATTAN—10 MILES. "Does this thing go any . . ulp . . . faster?"

The car leaped forward.

To his credit, Joe kept his eyes bravely open as Spud yanked the car in and out of high-speed traffic, snaking through holes that hadn't appeared to be there and doing unspeakable things to the Buick's transmission. But Joe was almost—almost—grateful when the sound of an ululating siren became audible over the snarling horns and screaming brakes.

Spud glanced in the mirror, located the whirling gumball machine in the rear-view mirror, and groaned aloud. "Just our luck! The cops—and us with only five bucks between us. Twelve years old, no license, a stolen car, a half a fat guy in a dress—cripes, even fifty bucks'd be cutting it close." Thinking furiously, he pulled over and parked on the grass, beneath a hellishly bright highway light. "Maybe I can go back and talk to them before they see you," he said to Joe, and began to get out.

"Wait, Spud!" Joe said urgently. He snatched a handful of cigarette butts from the ashtray, smeared black grime on Spud's upper lip. "There. Now you look maybe sixteen."

Spud grinned. "You're okay, Joe." He got out.

Twenty feet behind them, Patrolman Vitelli turned to his partner. "Freaks," he said happily. "Kids. Probably clouted the car, no license. Let me have it."

"Don't take a cent less than seventy-five," Patrolman Duffy advised.

"I dunno, Pat. They don't look like they got more than fifty to me."

"Well, all right," Duffy grumbled. "But I want an ounce of whatever they're smokin'. We're running low."

Vitelli nodded and got out of the black and white, one hand on his pistol. Spud met him halfway, and a certain lengthy ritual dialogue was held.

"Five bucks!" Vitelli roared. "You must be outa your mind."

"I wish I was," Spud said fervently. "Honest to God, it's all I got."

"How about your friend?" Vitelli said, and started for the Buick, which sat clearly illuminated in the pool of light beneath the arc-light.

"He's stone broke," Spud said hastily. "I'm takin' him to Bellevue—he thinks he may have leprosy."

Vitelli pulled up short with one hand on the trunk. "You got a license and registration?" he growled.

Spud's heart sank. "I . . ."

Vitelli nodded. "All right, buddy. Let's open the trunk."

Spud's heart bounced off his shoes and rocketed back up, lodging behind his palate. Seeing his reaction, Vitelli looked down at the trunk, noticing for the first time the odd nature of its fastening. He tugged experimentally, flimsy fabric parted, and the trunk lid rose.

Blinking at the light, the blond girl sat up stiffly, a muddy treadprint on her . . . person.

The air filled with the sound of screeching brakes.

Vitelli staggered back as if he'd been slapped with a sandbag. He looked from the girl to Spud to the girl to Spud, and his eyes narrowed.

"Oh, boy," he said softly. "Oh boy." He unholstered his gun.

"Look, officer, I can explain," Spud said without the least shred of conviction.

"Hey," said the blond girl, clearly dazed.

"Holy shit," said Duffy in the squad car.

"Excuse me," said Joe, getting out of the Buick.

Both cops gasped as they caught sight of him, and Vitelli began to shake his head slowly. Seeing their expressions, the girl raised up onto her knees and peered around the trunk lid, completing the task of converting what had been three lanes of rushing traffic into a goggle-eyed parking lot.

"My dress," she yelped.

Koziack stood beside the Buick a little uncertainly, searching for words in all the likely places. "Oh shit," he said at last, and began to pull the dress over his head, removing the derby. "Pleistocene, here I come."

Vitelli froze. The gun dropped from his nerveless fingers; the hand stayed before him, index finger crooked.

"Tony," came a shaky voice from the squad car, "forget the ounce."

Spud examined the glaze in Vitelli's eyes and bolted for the car. "Come on," he screamed at Joe. The girl barely (I'm sorry, really) managed to jump from the trunk before the car sprang forward like a plane trying to outrun a bullet, lurching off the shoulder in front of a ten-mile traffic pileup that showed no slightest sign of beginning to start up again.

Behind them Vitelli still stood like a statue, imaginary gun still pointing at where Joe had been standing. Tears leaked from his unblinking eyes.

As the girl stared around her with widening eyes, car doors began to open.

Spud was thoroughly spooked, but he relaxed a good deal when the toll-booth attendant at the Brooklyn Bridge failed to show any interest in a twelve-year-old driving a car with the trunk wide open. Joe had the dress folded over where his lap should have been, and the attendant only changed the five and went back to his egg salad sandwich without comment.

"Where are we going?" Spud asked, speaking for the first time since they had left the two policemen and the girl behind.

Joe named a midtown address in the Forties.

"Great. How're we gonna get you from the car into the place?"

Joe chuckled. "Hey, Spud—this may be 1976, but Manhattan is Manhattan. Nobody'll notice a thing."

"Yeah, I guess you're right. What do you figure to do?"

Joe's grin atrophied. "Jeez, I dunno. Get the belt fixed first—I ain't thought about after that."

Spud snorted. "Joe, I think you're a good guy and I'm your pal, but if you didn't have a roof on your mouth, you'd blow your derby off every time you hiccuped. Look, it's simple: you get the belt fixed, you get both halves of you back together, and it's maybe ten o'clock, right?"

"If those goniffs at the dealership don't take too long fixin' the belt," Joe agreed.

"So you give me the insurance money, and use the belt to go a few months ahead. By that time, with the Series and the Bowl games and maybe a little Olympics action, we can split, say, fifty grand. You take your half and take the time-belt back to the moment your legs left 2007, at 10:01. You buy your wife a time-belt first thing in the morning and you're both safe."

"Sounds great," Joe said a little slowly, "but . . . uh . . ."

Spud glanced at him irritably. "What's wrong with it?" he demanded.

"I don't want you should be offended, Spud. I mean, you're obviously a tough, smart little guy, but . . ."

"Spit it out!"

"Spud, there is no way in the world a twelve-year-old kid is gonna take fifty grand from the bookies and keep it." Joe shrugged apologetically. "I'm sorry, but you know I'm right."

Spud grimaced and banged the wheel with his fist. "I'll go to a *lot* of bookies," he began.

"Spud, Spud, you get into that bracket, at your age, the word has just gotta spread. You *know* that."

The boy jammed on the brakes for a traffic light and swore. "Dammit, you're right."

Joe slumped sadly in his seat. "And I can't do it myself. If I get caught bettin' on sports events of the past myself, it's the Pleistocene for me."

Spud stared, astounded. "Then how did you figure to accomplish *any*thing?"

"Well . . ." Joe looked embarrassed. "I guess I thought I'd find some guy I could trust. I didn't think he'd be . . . so young."

"A grownup you can *trust*? Joe, you really are a moron."

"Well. I didn't have no choice, frag it. Besides, it might still work. How much do you think you *could* score, say, on one big event like the Series, if you hustled all the books you could get to?"

Twenty thousand, Spud thought, but he said nothing.

Joe had been right: the sight of half a fat man being dragged across the sidewalk by a twelve-year-old with ashes on his upper lip aroused no reaction at all in midtown Manhattan on a Friday night. One out-of-towner on his way to the theater blinked a few times, but his attention was distracted almost immediately by a midget in a gorilla suit, wearing a sandwich sign advertising an off-off-off-Broadway play about bestiality. Spud and Joe reached their destination without commotion, a glass door in a group of six by which one entered various sections of a single building, like a thief seeking the correct route to the Sarcophagus Room of Tut's Tomb. The one they chose was labeled, "Breadbody & McTwee, Importers," and opened on a tall stairway. Spud left Joe at the foot of the stairs and went to fetch assistance. Shortly he came back down with a moronic-looking pimply teen-ager in dirty green coveralls, "Dinny" written in red lace on his breast pocket.

"Be goddamn," Dinny said with what Joe felt was excessive amusement. "Never seen anything like it. I thought this kid was nuts. Come on, let's go." Chuckling to himself, he helped Spud haul Joe upstairs to the shop. They brought him into a smallish room filled with oscilloscopes, signal generators, computer terminals, assorted unidentifiable hardware, tools, spare parts, beer cans, as-yet unpublished issues of *Playboy* and *Analog*, overflowing ashtrays, a muted radio, and a cheap desk piled with carbon copies of God only knew what. Dinny sat on a cigarette-scarred stool, still chuckling, and pulled down a reference book from an overhead shelf. He chewed gum and picked at his pimples as he thumbed through it, as though to demonstrate that he could do all three at once. It was clearly his showpiece. At last he looked up, shreds of gum decorating his grin, and nodded to Joe.

"If it's what I t'ink it is," he pronounced, "I c'n fix it. Got yer warranty papers?"

Joe nodded briefly, retrieved them from a compartment in the time-belt and handed them over. "How long will it take?"

"Take it easy," Dinny said unresponsively, and began studying the papers like an orangutan inspecting the Magna Charta. Joe curbed his impatience with a visible effort and rummaged in a nearby ashtray, selecting the longest butt he could find.

"Joe," Spud whispered, "how come that goof is the only one here?"

"Whaddya expect at nine thirty on a Friday night, the regional manager?" Joe whispered back savagely.

"I hope he knows what he's doing."

"Me too, but I can't wait for somebody better, dammit. Alice is in danger, and my legs've been using up my time for me back there. Besides, I've had to piss for the last hour-and-a-half."

Spud nodded grimly and selected a butt of his own. They smoked for what seemed like an interminable time in silence broken only by the rustling of paper and the sound of Dinny's pimples popping.

"Awright," the mechanic said at last, "the warranty's still good. Lucky you didn't come ta me a week from now."

"The speed you're goin', maybe I have," Joe snapped. "Come on, come on, will ya? Get me my legs back— I ain't got all night."

"Take it easy," Dinny said with infuriating glee. "You'll get your legs back. Just relax. Come on over inna light." Moving with sadistic slowness, he acquired a device that seemed something like a handheld fluoroscope with a six-inch screen, and began running it around the belt. He stopped, gazed at the screen for a full ten seconds, and sucked his teeth.

"Sorry, mister," he drawled, straightening up and grinning. "I can't help you."

"What the hell are you talkin' about?" Joe roared.

"Somebody tampered with this belt, tried to jinx the override cutout so they could visit some Interdicted Period—probably wanted to see the Crucifixion or some other event that a vested-interest group get declared Off-Limits. I bet that's why it don't work right. It takes a specialist to work on one of these, you know." He smiled proudly, pleased with the last sentence.

"So you can't fix it?" Koziack groaned.

"Maybe yes, maybe no, but I ain't gonna try 'less I see some cash. That belt's been tampered with," Dinny said, relishing the moment. "The warranty's void."

Joe howled like a gutshot buffalo, and stepped forward. His meaty right fist traveled six inches from his shoulder, caught Dinny full in the mouth and dropped him in his tracks, popping the mechanic's upper lip and three pimples. "I'd stomp on ya if I could, ya smart-ass mugger-hugger," Joe roared down at the unconscious Dinny. "Think you're funny!"

"Easy, Joe," Spud yelled. "Don't get excited. We gotta *do* something."

"What the hell *can* we do?" Joe cried despairingly. "That crumb is the only mechanic in a hundred miles—we'll never get to the next one in time, and we haven't got a prayer anyway with four dollars and change. Crummy pap-lapper, I oughta . . . oh *damn* it." He began to cry.

"Hey, Joe," Spud protested, flustered beyond measure at seeing a sober grownup cry, "come on, take it easy. Come on now, cut it out." Joe, his face in his hands, shook his head and kept on sobbing.

Spud thought furiously, and suddenly a light dawned and he was filled with a strange prescience, a déjà vu kind of certainty that startled him with its intensity. He wasted no time examining it. Stepping close to Joe, he bent at the waist, swung from the hip, and kicked the belt as hard as he could, squarely on the spot Dinny had last examined. A sob became a startled yell—

—and Joe's fat legs appeared beneath him, growing downward from the belt like tubers.

"What the hell did you kick me for?" Joe demanded, glaring indignantly at Spud. "What'd I do to you?"

Spud pointed.

Joe looked down. "Wa-HOO!!" he shouted gleefully. "You *did* it, Spud, I got my legs back! Oh, Spud, baby, you're beautiful, *I got my legs back!*" He began to caper around the room in spontaneous improvised goat-dance, knocking equipment crashing in all directions, and Spud danced with him, laughing and whooping and for the first time in this story looking his age. Together they careened like an improbable vaudeville team, the big fat man and the mustached midget, howling like fools.

At last they subsided, and Joe sat down to catch his breath. "Woo-ee," he panted, "what a break. Hey, Spud, I really gotta thank you, honest to God. Look, I been thinkin'—you can't make enough from the bookies for both of us without stickin' your neck way out. So the hell with that, see? I'll give you the Series winner like I promised, but you keep all the dough. I'll figure out some other way to get the scratch—with the belt workin' again it shouldn't be too hard."

Spud laughed and shook his head. "Thanks, Joe," he said. "That's really nice of you, and I appreciate it— but 'figuring out' isn't exactly your strong suit. Besides, I've been doing some thinking too. If I won fifty bucks shooting pool, that'd make me happy—I'd be proud, I'd've earned it. But to make twenty thousand on a fixed game with no gamble at all—that's no kick. You need the money—you take it, just like we planned. I'll see the bookies tonight."

"But you earned it, kid," Joe said in bewilderment. "You went through a lotta work to get me here, and you fixed the belt."

"That's all right," Spud insisted. "I don't want money—but there's one thing you *can* do for me."

"Anything," Joe agreed. "As soon as I take a piss."

Three hours later, having ditched the car and visited the home of "Odds" Evenwright, where he placed a large bet on a certain ball-club, Spud arrived home to find precisely what he had expected:

His mother, awesomely drunk and madder than hell, sitting next to the pool table on which his personal cue and balls still rested, waiting for him to come home.

"Hi, Mom," he said cheerfully as he entered the living room, and braced himself. With a cry of alcoholic fury, Mrs. Flynn lurched from her chair and began to close on him.

Then she pulled up short, realizing belatedly that her son was accompanied by a stranger. For a moment, old reflex manners nearly took hold, but the drink was upon her and her Irish was up. "Are you the tramp who's been teachin' my Clarence to shoot pool, you tramp?" she screeched, shaking her fist and very nearly capsizing with the effort. "You fat bum, are you the one'sh been corrupting my boy?"

"Not me," Joe said politely, and disappeared.

"They ran out of pink elephants," he explained earnestly, reappearing three feet to the left and vanishing again.

"So I came instead," he went on from six feet to the right.

"Which is anyway novel," he finished from behind her, disappeared one last time and reappeared with his nose an inch from hers. Her eyes crossed, kept on crossing, and she went down like a felled tree, landing with the boneless grace of the totally stoned.

Spud giggled, and it was not an unsympathetic giggle. "Thanks, Joe," he said, slapping the $\frac{\text{fat man}}{1}$ on the back. "You've done me a big favor."

"Glad I could help, kid," Joe said, putting his own arm around the boy. "It must be tough to have a juicer for an old lady."

"Don't worry, Joe," Spud said, feeling that same unexplainable certainty he had felt at the time-belt repair shop. "Somehow I've got a feeling Mom has taken her last drink."

Joe nodded happily. "I'll be back after the Series," he said, "and we can always try a second treatment."

"Okay, but we won't have to. Now get out of here and get back to your wife—it's late."

Joe nodded again. "Sure thing, Spud." He stuck out his hand. "Thanks for everything, pal—I couldn't have made it without you. See you in a couple o' weeks, and then, who knows—Alice an' I might just decide this era's the one we want to settle down in."

"Not if you're smart," Spud said wryly.

"Well, in that case, maybe I'll be seein' ya again sometime," Joe pointed out. He reached down, made an adjustment on the time-belt, waved good-bye and vanished.

Or nearly. A pair of fat legs still stood in the living room, topped by the time-belt. As Spud stared, one of the legs stamped its feet in frustration and fury.

Sighing, Spud moved forward to kick the damned thing again.

TIDBIT: two puns

Rhythms and 'Rithms (COURTESY OF SUSI STEFL)

The evolutionary forebears of all the poisonous snakes now living were a pair of adders who managed to book passage with Noah during the Flood. They were not permitted aboard the Ark itself, however. The little dinghy full of skunks traditionally seen trailing from the stern of the Ark in cartoons was in fact one

of a whole flotilla of makeshift rafts on which animals noxious to man were towed to Ararat. The one the snakes drew was a crude table carved from a single, massive log.

It perfectly suited their dimensions, of course, but its enforced restriction of position had an inevitable consequence. On the third day of the voyage, Noah, who was teaching a mathematics class on the fantail to alleviate shipboard ennui, noticed that a number of his students were not paying attention; they kept glancing furtively over the stern and giggling. Noah soon spied the source of their amusement: the two snakes were making enthusiastic and acrobatic love.

"There, you see?" cried Noah, seizing a chance to underscore a lesson. "Even adders can multiply on a log table!"

The Shamin' of the Shaman

Running Nose, the toughest Indian in the tribe, showed up for sick call at the medicine man's tent, complaining of persistent diarrhea. Leaping Fees, the withered shaman, gave him a single strip of rawhide by way of prescription. "Bite an inch off this each day," he intoned. "Chew it well and swallow. The hide is cured; you will be too. That'll be three buffaloes."

Nose went away well pleased, although three buffaloes cleaned him out. But a week later he returned in a Fury (or perhaps it was a Corvair . . .), one hand on his belly and the other on his tomahawk. Plainly, Running Nose was still running.

Leaping Fees sat down at once and began to sing his death song. And its first line was: "The thong is ended/but the malady lingers on . . ."

3
TOO SOON WE GROW OLD

The first awakening was awful, and she enjoyed it.

She was naked and terribly cold. She was in a plastic coffin from whose walls grew wrinkled plastic arms with gnarled plastic hands that did things to her. Most of the things hurt dreadfully—but they were all physical hurts. Her soul was conscious only of an almost terrifying sense of relief. Until you have had your neck and shoulders rubbed out for the first time, you can have no conception of how tightly bunched they were. Tension can only be fully appreciated in its release. To her mind came a vivid association from long decades past: her first orgasm. A shudder passed over her body.

A voice spoke, in a language unknown to her. Even allowing for the sound-deadening coffin walls, it sounded distant.

Eyes appeared over hers, through a transparent panel she had failed to see since it had showed only a ceiling the same color as the coffin's interior. She refocused. The face was masked and capped in white, the eyes pouched in wrinkles. He said something incomprehensible, apparently in reply to the first voice.

"Hi, Doc," she shouted, finding her voice oddly squeaky in the high-helium atmosphere of the cryogenic capsule. *"I made it!"* She found that she was grinning.

He started, and moved from view. One of the plastic hands did something to her left bicep, and she felt her hurts slipping away—but not her joy. *I knew I could beat it,* she thought just before consciousness faded, and then she dreamed of the day her victory had begun.

She was not at all sure just why she had consented to the interview. She had rejected them for over twenty years, on an impulse so consistent that it had never seemed to call for examination. To understand why she had granted this one would, it seemed to her, call for twenty years' worth of spade-work—it was simpler to posit that impulse had merely changed its sign, from negative to positive.

Yet, although she relied implicitly on the automatic pilot which had made the decision for her, she found apprehension mounting within her as the appointed day led her inexorably to the appointed time. An hour before the interviewer was due, she found herself examining a capsule of an obscure and quite illegal tranquilizer, one which had not even filtered down to street level yet. It was called Alpha, according to her source, and he claimed it was preternaturally effective. But she hesitated—he had said something about it tending to suppress *all* the censors, something about it being a kind of mild truth drug. She turned the capsule end on end in her palm, three times.

The hell with it, she decided. *This is the true measure of my wealth: I can even afford to be honest with an interviewer.* The realization elated her. *Besides,* she afterthought, *I can always buy the network if I have to.*

She washed the capsule down with twice-distilled water.

The lights were not as blinding as she had expected. In fact, none of the external irritations she had anticipated materialized—not even the obtrusive presence of a cameraman. The holocamera was *not* entirely automatic, for newstaping is an art (with a powerful union)—but its operator was nearly a hundred miles away in the network's headquarters, present only by inference. She was simply sitting in her own familiar living room, conversing with a perfect stranger whose profession it was to seem an old and understanding friend. Although she had never seen his show—she never watched 3V—she decided he was one of the best in his field. Or was that the drug? In any case, they went from Ms. Hammond and Mr. Hold to Diana and Owen in what was, for her, a remarkably short time. As she realized this, alarm made one last attempt to take over her controls, but failed.

". . . clearly done a number of remarkable things with your half-century, Diana," Hold was saying with obvious sincerity. "Today it is no longer inconceivable for a woman to become wealthy by her own efforts in the economic marketplace—but you began your fantastic career in an age when as a rule, only men had such opportunities. In fact, you've done as much as, perhaps more than anyone to bring our society out of that restrictive phase."

The words warmed her. "Oh," she said lightly. "It's not a difficult trick to become terribly rich. 'All it takes is a lifetime of devotion.'"

"I'm familiar with the quote," Hold agreed. "All the same, it must have been an incredibly difficult, demanding task to carve yourself a navigable path where none existed. And so perhaps the foremost question in my mind is, why?"

"I beg your pardon?"

"Why did you choose the life you have led? What was your motivation for this lifetime of devotion?"

"Because," she said almost automatically, "given the nature of the world I found myself in, it seemed the most sensible, the most mature . . . the most grown-up thing to do."

"I'm not sure I understand," Hold said, and he *was* the best in his field, because he had the rare gift of listening totally, of conveying by his utter attention to her every gesture and nuance his eagerness to *understand*. Since everyone knows that to understand is to forgive all, no one who genuinely wishes to understand can be an enemy—can they?—and so she found herself explaining to another the agony that had been her childhood.

". . . and so with Father dead and five girls to raise, Mother entered the business world. She had to— Father's insurance company flatly refused to pay. They claimed it was clearly a suicide, and three judges agreed. There was still a sizable estate, of course, but after the deduction of lawyers' fees and nonrefundable losing bids on three judges, it wasn't enough to provide for all six of us for very long. So Mother converted it all to capital and tried to become a business woman, about the time I was twelve. In today's world she might have succeeded—but she was terribly ignorant and naive. Father's inherited wealth had sheltered her as effectively as it had him. The only people who paid her any attention, let alone respect, were the sharks, and they had picked her clean by the time I was twenty. That was . . . let me see . . . 1965 or 66.

"And so it was up to me, the eldest. Mother had gotten clever in the final extremity: no one ever called her death anything but accidental. But even so, the inheritance I received was almost nominal.

"But it was enough, for me and for my sisters."

"Clearly," Hold agreed. "Then you would say your initial motivation was to provide for yourself and your sisters."

"More for them than for myself," she said, and was gratified to hear herself say so. "Mother had passed on to me her own overwhelming sense of responsibility. As matter of fact, my own strongest interest was in music. But I knew I could never provide five siblings on a musician's wages, and so I put all that away and buckled down."

"You must be deeply happy, then," he said, "to have so thoroughly realized your life's ambition."

And she surprised herself. "No. No, I can't say that I am."

His face, his posture, his body-language all expressed his puzzlement.

"Perhaps," she said slowly, hearing the words only as they came from her mouth. "perhaps one's life ambition oughtn't to be something that can be achieved. Because what do you do *then?* Perhaps one's life ambition should be something that will always need to be worked at."

"But surely you're a long way from retirement?"

"Medically, yes," she agreed. "My doctors tell me I can look forward to at least twenty more years of excellent health. Surely I can contrive to push mountains of money back and forth for that long. But *why?* I have already achieved total security. If I were to seal myself up in my bathroom, my fortune would continue to grow—it has passed the critical point for self-sustaining reaction. And all my sisters are now independent, one way or another.

"I have been . . . uneasy, for months now, discontent in a way I could not explain to myself. But I see it now: I've achieved all I set out to do. No wonder I've been so . . ." She broke off and lapsed into deep thought, utterly unaware of the holocamera.

"But surely," Hold began again, "there are other goals you can turn your attention to now."

"What goals?" she asked, honestly curious.

"Er . . . well, the classic ones, of course," he said. "That is . . . well, to make the world a better place . . ."

"Owen," she said. "I confess that after half a century of living, I haven't the faintest notion of how to make the world a better place. I wouldn't know where to begin."

"Well, then, to leave something better behind for . . ."

"For posterity?" she finished. "Look at me. *I'm fifty-four years old.*"

Hold was silent.

"In fact, that may be the single craziest thing about this society," she said, her voice rising. "We're best prepared to bear children, biologically, in our teens—and we're best able to raise them, socially and economically, in our middle and later years. For the first time in my life, my responsibilities have eased to the point where I can consider children of my own—and now I'm too old to *have* them." The camera unobtrusively tracked her as she rose and paced around the spacious living room. "I've been a surrogate mother for years, and now I'll never be a *real* one."

"But, Diana," Hold cut her off, "surely parenthood is not the only form of immortality available to someone of your . . ."

"You don't understand," she cried. "I don't want immortality, even by proxy. I want *children*. Babies, of my own, to cherish and teach and raise. All my life I've sublimated my maternal drive, to feed and clothe and house my sisters. Now that's ended—and it was never really enough to begin with. Oh, why didn't . . ." She flung out an arm, and the very theatricality of the gesture reminded her all at once that she was being recorded. She dropped the arm and turned away from the camera in confusion. "Owen . . . Mr. Hold, I must ask you to leave now. I'm sorry but this interview was a mistake."

With a total absence of dismay, Hold rose fluidly from the powered armchair and faced her squarely. Perhaps it was coincidental that this presented his best profile to the camera. "You know yourself better now, Diana," he said. "That may sting, but I hardly think it can be a mistake."

"If you're trapped in a canyon, aren't you better off not knowing?" she asked bitterly.

"Subconsciously you knew all along," he countered. "At least now you're facing the knowledge. What you *know* can't be cured, you know you can endure."

She examined her fingertips. "Perhaps you're right," she said at last. "Good day, Mr. Hold."

"Good-bye, Diana."

He collapsed the camera and left, looking rather smug.

A long time later, seated on a couch which had cost the equivalent of her father's total worth at the time of his death, she said to the empty air, ". . . but I never have been a quitter."

And after the sun had come up, she called her local Cold Sleep center, made an appointment to speak with its director, and then called her attorney.

The second awakening was much better, and she did not enjoy it nearly as much.

Objectively, she should have. She no longer hurt anywhere that she could detect, and the bed—she corrected herself—the artifact on which she was half-sitting was the next best thing to an upholstered womb for comfort. She was alone in an apparently soundproof hospital room, in which the lighting was soft and indirect. She was neither hungry nor unhungry, neither weary nor restless.

But she was uneasy, as though in the back halls of her mind there faintly yammered an alarm bell she could not locate, an alarm clock she could not shut off. It was an unreasoned conviction that *something is wrong*. Unreasoned—was it therefore unreasonable?

That called for a second opinion.

Before she had given herself up to cryogenic sleep, she had firmly instructed herself not to be childishly startled by unfamiliar gadgetry when she woke. All the same, she was startled to learn that her nurse-call buzzer was (a) cordless, (b) conveniently accessible, and (c) nonspring-loaded, so that it could be thumbed without effort. It was not the technology that was startling—she realized that such technology had existed in her own time—it was the thoughtful compassion which had opted to use technology for patient-comfort. *Maybe they've repealed Murphy's Law*, she thought wildly, and giggled. *Now there's a dangerous vision for you.*

She was even more profoundly startled to learn that the *other* end of the process had been equally improved: her summons was answered at once. A tall, quite aged man with a mane of white hair swept aside the curtain at the far corner (the room couldn't be soundproof, then. Could it?) and stepped into the room. His clothing startled her again. She was somewhat used to the notion of purely ornamental, rather than functional, clothing—but to her mind, "ornament" involved not-quite-concealing the genitals. Embarrassed, and therefore furious with herself, she transferred her gaze to his face, and felt her emotional turmoil fade, leaving only that original undefined unease like a single rotting stump protruding from a vast tranquil lake.

His mouth was couched in strong wrinkles that spoke of frequent laughter and tears, and his eyes were a clear warm blue beneath magnificent white eyebrows. She was . . . not captured . . . held by those eyes; to meet them was to be stroked by strong, healing hands, hands that gently probed and learned and comforted. They made her smile involuntarily, and his answering smile was a kind of benediction, a closing of a circuit between them.

And then those eyes seemed to see the rotting stump; the great white brow frowned mightily. "What's the matter, Diana!"

She could not frame the words; they simply spilled out. *"How much time has passed?"*

Comprehension seemed to dawn, yet the frown deepened. "Even more than you stipulated," he said carefully.

She knew, somehow, that he would not lie, and tried to relax. It did not entirely work. *I've achieved what I set out to*, she thought, *but there's a catch of some kind somewhere. Now how do I know that?* Then she thought, *More important, how did he know that?*

"Who are you?" she asked.

He was perceptive enough to guess which question she had asked with those words. "I am Caleb," he said. "You've evidently guessed that I'm to be your Orientator."

"I'm fairly good at anticipating the obvious," she said proudly. "It was inevitable that someone would have to fill me in on current conventions, show me how to recognize the ladies' room and so forth."

He laughed aloud. "I'm afraid that by 'anticipating the obvious,' you mean straight-line extrapolation of what you were already accustomed to. That's going to cause you problems."

"Explain," she said, wondering if she should take offense at his laughter.

"Well, for a start, I can't show you how to find a ladies' room."

"Eh?"

"I can show you how to find a public toilet."

She registered confusion.

He smiled tolerantly. "Come now—you're obviously quite intelligent. What does your term imply that mine does not?"

She thought a moment. "Oh." She reddened. "*Oh.*" She went on thinking, and he waited patiently. "I suppose that makes sense. Earth must be too crowded by now to duplicate facilities without good reason."

He laughed aloud again, and this time she tried to take offense. Since Caleb was not offering any, she failed. "There you go again. You'll simply have to stop assuming that this is your world with tailfins on it. It isn't, you know."

"Will you explain my error?" she asked, battling her own irritation.

"It's not that we needed to stop excreting in secret—it's that we *stopped* needing to *do* so."

She thought that over very carefully indeed, and again Caleb waited with infinite patience. He clearly understood that she wanted to work out as much of it as possible for herself—in order to deny that this strange new world was quite terrifying.

"Another question," she said finally. "When was the last war?"

His smile was more than approving—it was congratulatory, quite personally pleased. "Well," he said, "last night a few thousand of us had one hell of an argument over next year's crop program. Some of the younger folk got quite exasperated. But if you mean physical violence, deliberate damage . . . well, I'm a historian, so I could give you the precise date. But if you were to step out into the hall and ask someone, you'd probably get a blank stare. Does that answer your question?"

"Yes," she said slowly. "You're telling me that we've . . . that the race has actually . . ." She paused, found the word. "Actually grown up."

"We like to think that we're adjusting well to adolescence," he said. "Of course, that implies the same sort of extrapolation you've been trying to use—but that's the best *we* can do, too."

"You are wondrously tactful, Caleb," she said. "But dammit, my whole life till now has been based on extrapolation."

"Oh, on a short time-scale it works just fine," Caleb agreed. "But over a long range, it works only as hindsight. It's a matter of locating the really significant data from which to extrapolate. An extrapolator in the early 1900s might have been aware that a man named Ford had invented a mechanical horse—but how could that observer have guessed how *much* significance that should have in his projections? All the seeds of today were present in your world, and you were almost certainly aware of them. But if I hand you a thousand seeds, most of which are strange new hybrids, how are you to know which will be weeds and which mighty trees?"

"I understand," she said, "but I must admit I find the idea disturbing."

"Of course," he said gently. "We all like to think ourselves such imaginative navigators that no new twist in the river can startle us. The one thing that every Awakened Sleeper finds *most* surprising is the depth of his own surprise. The fun in all stories is trying to guess what happens next, and we like to feel that if we fail, it's either because we didn't try hard enough or because the author cheated. God is a much more talented author than that—thank God."

"I suppose you're right," she agreed. "All right, what were the seeds—the data I overlooked?"

"The biggest part was, as far as I can tell, right under your nose. The spiritual renaissance in North America was already well under way in your time."

Her jaw dropped in honest astonishment. "Do you mean to tell me that all that divine mumbo-jumbo, all those crackpot holy men, actually produced anything?"

"The very success of such transparent charlatans proved that they were filling a deep and urgent need. When the so-called 'science' of psychology collapsed under the weight of its own flawed postulates, its more sincere followers perforce turned their attention toward spirituality. Over the ensuing decades, this culminated in the creation of the first self-consistent code of ethics—one that didn't depend on a white-bearded know-it-all with thunderbolts up his sleeve to enforce it. It didn't have to be enforced. When completed, it was as self-evidently superior to anything that had gone before as the assembly line was in its time. It sold itself. Behaviorally-determined helplessness may be a dandy rationalization—but it isn't any *fun*.

"At more or less the same time, there was a widespread boom in use of a new drug called Alpha . . . why are you frowning?"

"I'm familiar with Alpha," she said sourly. "Salvation by drug addiction—that sounds just great."

"You misunderstand," he said gently. "It's not that the drug is addictive. Happens it's not. It's Truth that's addictive."

"Go on," she said, plainly not convinced.

"An interest in spirituality, combined with volitional control of rationalization, led inevitably to the first clear distinguishing between pleasure and joy. Then came the first rigorous definition of sharing, and the rest followed logically, Shared joy is increased; shared pain lessened. Axiomatic."

"But I knew that," she cried, and caught herself.

"If so," he said with gentle sadness, "then—as I have just said—you did not know how much significance to assign the awareness. What clearer proof is there than your presence here—than the inescapable fact that you used a much greater proportion of the world's resources than you deserved, specifically to remove yourself from all possibility of sharing with anyone you knew?"

"Wait a damn minute," she snapped. "I *earned* my fortune, and furthermore . . ."

"It is impossible," he interrupted, "to *earn* more than you can use—you can only acquire it."

". . . and *furthermore*," she insisted, "I risked my life and health on the wild gamble that your age would Awaken me, specifically so that I *could* share— share my life and my experience with children."

"Whose children?" he asked softly.

She blew up. "You garrulous old fool, what in the HELL was the point in considering that until it was a physical possibility? How do I know whose children? Perhaps I'll have myself artificially inseminated, perhaps I'll have me a virgin birth, what business is it of yours whose children?"

"Am I not my sister's keeper?" he asked, unmoved by the violence of her rage. "Admit it, Diana: you *have* considered the matter, even if only subconsciously. And those flip, off-the-top-of-your-head suggestions are all you've come up with. Sharing the job of parenthood might just be one of the most exciting challenges of your life—but what you really want is only to re-create a game you already know how to win: raising images of yourself by yourself."

Implied insult could enrage her but when she felt herself directly attacked she invariably became calm and cold. The anger left her features, and her voice was "only" impersonal. "You make it sound easy. Being father and mother both."

"Easy?" he said softly. "It cannot be done—save poorly, when there is no alternative. 'Poorly,' of course, is a relative term. Fate gave you that very burden to shoulder, and you did magnificently—from the records I have, it appears that none of your sisters turned out significantly more neurotic than you."

"Except Mary," she said bitterly.

"There is no reason to believe that you could have prevented her tragedy," he said. "I repeat: given what you had to work with, you did splendidly. But if you persist in trying to repeat the task with no more than you had to work with then, you will end in sorrow."

"It would be challenging," she said.

"If it's challenge you want," he said in exasperation, "then why don't you try the one that occupies our attention these days?"

"And that is?"

"Raising the sanest children that it is within our power to raise. It's the major thrust of current social concern, and the only ethical approach to procreation. How else are we to grow up than by growing ourselves up?"

"And how do you do that?" she asked, intrigued in spite of herself.

He tugged at the ends of his snow-white hair. "Well, some of it I can't explain to you until you've learned to talk—in our speech, I mean. I like this old tongue, but it's next to impossible to think coherently in it. But one of the basic concepts you already know.

"I reviewed a copy of that final interview you gave,

to that man with the unbearably cute name. Owen B. Hold, that was it, of *Lo And Behold*. It's a rather famous tape, you know: you made a big splash in the media when you opted for Cold Sleep. Richard Corey has always been a popular image.

"And in that interview you raised one of the central problems of your age: the biological incongruity by which humans lost the capacity for reproduction at just about the time they were acquiring the experiential wisdom to raise children properly. People were forced to raise children, if at all, during the most agonizingly confused time in their lives, and by the time they *had* achieved any stability or 'common' sense, they tended to drop dead.

"Technology gave us the first phase of the solution: rejuvenation treatments were developed which restored fertility and vigor to the aged. The second phase came when the race abandoned technology—that is, clumsy and dangerous technological means of birth control—and learned how to make conception an act of the will. The ability was always there, locked in that eighty-five percent of the human brain for which your era could find no use. Its development was a function of increased self-knowledge.

"The two breakthroughs, combined, solved the problem, by encouraging humans not to reproduce until they were truly prepared to. The effects of this change were profound."

He broke off, then, for she was clearly no longer listening. *That's it*, she thought dizzily, *that's the final confirmation, he's just told me I can do it, so why am I still* sure *there's a catch to it somewhere? There's too much happening at once, I can't think straight, but something's wrong and I don't know what it is.* "Have you supermen figured out what a hunch is?" she said aloud.

Apparently Caleb had the rare gift of moving without attracting attention, for he was now in a far corner of the room. He seemed to have caused the wall to extrude something like a small radar screen; his eye movements told her he was studying some display she could not see. When he spoke, his voice was grim. "That was known in your time. A hunch is a projection based on data you didn't know you possessed. Like the one that's been bothering you since I came into this room—the one that's been mystifying me for the same period."

She shook her head. "I've had this hunch since before I could possibly have had *any* data—from the moment I regained consciousness in this room."

"Except while actually in cryogenic stasis," he said, "no one is *ever* unconscious."

She started to argue, and then remembered the time she had been involved in a traffic accident on her way through South Carolina. She had spent a week in a coma, and awakened to find that she was speaking in a pronounced Southern drawl which mirrored that of her nurses. "Perhaps you're right," she conceded. "But what data could I have?"

He gestured to the screen. "This device monitors the room, so that if you should be found in the midst of some medical crisis, the Healers can study its beginnings. I've run it back, and I've found the problem."

A sudden increase in anxiety told her that Caleb was right. "*Well?*"

"A man with a habit of talking to himself did so as he was wheeling you in here from surgery and tuning your bed to you. This might not have mattered—save that he, too, is a recently Awakened Sleeper, who still thinks and speaks in his old tongue. Since that tongue is Old English, his mutterings upset your subconscious—and forced my hand. I hadn't wanted to go into this on your first post-Awakening conversation, Diana—but now I have no choice."

She tried to tense her shoulders, but the bed would not permit it. She settled for clamping her teeth together. "Let's get on with it."

"I'm afraid you *can't* have children—yet. Possibly never."

"But you said . . ."

"That even more time had passed than you stipulated, yes. Your instructions were to Awaken you as soon as it became medically possible for a woman of your age to bear optimally healthy children in safety. That condition obtains, and has for some time."

"And you're telling me I *can't*, even though it's medically possible." Her lip curled in a sneer. "I thought this Brave New World was too good to be true. Go on, Caleb—tell me more about your little Utopian tyranny."

"You misunderstand."

"I'll bet. So procreation is an 'act of the will,' eh? Just not mine."

"Diana, Diana! Yes, procreation is everyone's personal responsibility. But it requires *two* acts of the will."

"What?"

"I am not saying that you are forbidden to procreate. I'm saying that you won't find a male—or a clinic—willing to cooperate with you at this time."

"*Why not?*"

"Because," he said with genuine compassion, "you're not old enough."

Before shock gave way to true sleep, she became aware of Caleb again, realized that he had never left her side. He was holding her hand, stroking it as gently as a man removing ashes from a third-degree burn.

It was an enormous effort to speak, but she managed. "Will . . . will I . . ."

Caleb bent closer.

"Will I ever be old enough?" she whispered.

A faint smile came to his thin, old lips. "Perhaps," he said softly. "Barring accident, you will live at least another seventy years, years of youthful vigor. But I must warn you that, by our standards, you are a backward child."

"Hell with . . . hell with that. Only thing you've . . . got I haven't . . . s'healthier background."

"That is true."

"Jus' . . . watch me. Never was a quitter."

"I know," he said, his smile widening. "Your file told me that. That's why I overrruled my colleagues, and Awoke you. I think that you will find joy, Diana. It's right in front of you. It always was." He paused. "Didn't you say something in that interview about having always wanted to study music?"

"Maybe I'll . . . have time to try it now."

He radiated approval. "Excellent. A life's ambition ought to be something that will always need to be worked at."

Peace washed over her, in something too gentle to be called a wave. She felt sleep reaching for her. But as her vision faded, curiosity birthed one last question.

"How many . . . how many thousands of years . . . has it been?"

His grin was something that could be heard and felt.

"Less than a century."

TIDBIT: two songs

VALKYRIE RIDE

(words and music by Spider Robinson)

The clouds are black and heavy as they lie upon the earth
And the wind is howling madly at the moment of my birth
And the sky is split with lightning as I draw a trembling
  breath
And my mother realizes that her son was born for
  Death . . .

(chorus)
> And the Valkyrie rides and there's death at her side
> Her visage is graven in black
> For when Odin decides, then the Valkyrie rides
> And you know she will never turn back

I grow to bloody manhood and my name is called The
  Grim
And I steal for my amusement and I murder on a whim
And the women fright their children with the stories that
  they tell
Of the many sorry warriors my sword has sent to Hell

I lead a band of hillmen now, I lead them with my blade
And we rape and kill and pillage as we ply our bloody
  trade
And they say I am a demon, yes, and tremble all the while
And they say no man alive or dead has ever seen me smile
(chorus)

Six hundred sons of Loki now are ready to my side
And a multitude have faced us and a multitude have died
We have sacked the Seven Cities and the villages between
And I slay the king and court and to my men I give the
    queen

And the Sorcerers conspire, and against me set their hand
But I pray to Father Loki and I slay them to a man
Then my journey to my weirding place is finally begun
As I lead six hundred killers to the City of the Sun . . .

And they send a host against me, and we meet upon a
    plain
And the world will never see such awful slaughtering again

VALKYRIE RIDE

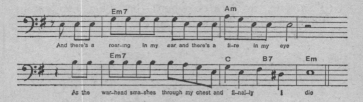

For they came with but a single thought: that even as
    they fell
An attacker of their city would precede them into Hell
(chorus)

And I fight with savage fury, and before me none can
    stand
And a score of mighty warriors have died before my hand
. . . When I see beyond the battle, flying high above the
    plain
A woman dressed in black, upon a horse without a
    name . . .

She raises in the saddle and she cocks a mighty lance
And the fighting stops around me as they catch her in their
    glance
And it seems to take forever for her arm to make the cast
As we wait to see which one of us is going to breath his
    last
And her arm is down and now the spear is flying straight
    and true
And we wait in silent terror for there's nothing else to do
And there's a roaring in my ear and there's a fire in my eye
As the warhead smashes through my chest and finally I die
        And the Valkyrie rides and there's death at her side
        Her visage is graven in black
        For when Odin decides, then the Valkyrie rides
        And you know she will never turn back . . .

(This song was written when I had been resident in Phinney's Cove, Nova Scotia, for approximately one week. I sing it still.)

FEED ME FIRE

(words and music by Spider Robinson)

Sittin' here thinkin' on a stormy old Monday
Watch the cat chewin' up a mouse
Wind is mighty cold whippin' off the Bay of Fundy
But it's toasty and warm here in the house
  And my friends have gone; I'm here alone; nothin'
   much to do
  But feed me fire pile her higher
  I reckon that'll do . . .

Sittin' here listenin' to my guitar singin'
A new Nova Scotia song
Sittin' here listenin' to my party line ringin'
One long, one short, one long
  Well that ain't my code I mighta knowed All I
   have to do
  Is feed me fire pile her higher
  That'll pull me through . . .

You know I can't remember when I ever *been* so happy
Happier than I can say
You know I used to feel older than my own grandpappy
But I'm gettin' younger every day . . .

Sittin' here listenin' to the cuckoo tickin'
I gotta say I'm feelin' good
Sittin' listenin' to my own guitar pickin'—
Wups!—believe I need a bit more wood
  Sip some o' this homebrew beer I like it here

And all I ever want to do
Is feed me fire   pile her higher   write a song or
   two
I reckon that'll do

FEED ME FIRE

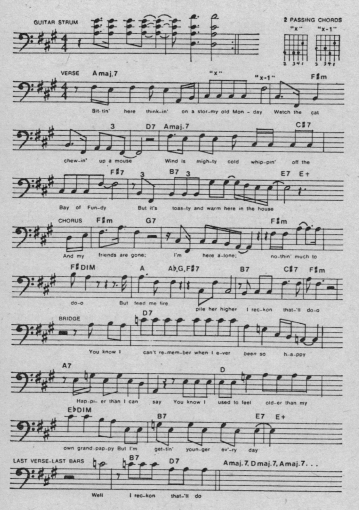

4
WHEN NO MAN PURSUETH

"Yes, indeed, m'boy," wheezed old Colonel Enderby-Thwaite, blinking at his cards, "if you want to get a real education before you get to Secundus, you've certainly picked the right way to go about it. Riding a tramp like *This Train* will teach you more about life with a capital 'L' than all the seminars and professors on Secundus, I daresay."

"Precisely my thought," Fleming agreed in what he hoped were mature tones. Although he faced the pot-bellied Colonel to whom he was teaching poker, he was aware in every nerve ending of the impossibly beautiful girl in blue seated across the passengers' lounge, at whom he had been sneaking glances ever since her arrival. In point of fact, he had booked passage on the *I.V. This Train* because it represented a saving of over a hundred credits; but before this girl he wished very desperately to appear worldly—or, if he could not pull that off, at least eager to be. The frequent glances at her were no help either, for she stared right back at him; and while she was *not* staring the sandy-haired student between the eyes, that rhymes with what she was staring between. Fleming racked his brains in vain for a means of introduction, even as he took the Colonel for two-credits-six.

"Not many realize what it's like out here," the Colonel rumbled on in the fond belief that Fleming was listening. "Fringe worlds. The incredible diversity. Ultimate solution to the minority opinion problem, actually. Someone's got a crackpot idea on how things should be run, give him a planet and let him try it out. Makes for some interesting planets. Why, over on Why Should I, they've actually done away with taxation. Except on a voluntary basis, of course—but if a politician has some project he thinks should be undertaken, he has to convince people to pay for it—and any project that people aren't willing to lay out cold cash for is scrapped. Only time will tell if it's viable, naturally, but it certainly is one of the most streamlined governments I've ever seen. And for centuries it was only a crackpot idea.

"Yes, m'boy, there's room out here on the Fringe for just about any kind of society. You don't see that sort of thing on the big passenger liners, nonstop jumps from Federation planet to Federation planet with their bland, homogeneous 'culture' to make them identical. Out here there's variety. You meet people who think differently, live differently than you. Stimulating."

"Sometimes," said the girl in blue in a voice like the mellowest of clarinets, "it can be very stimulating indeed."

This seemed to Fleming a clear-cut invitation to repartee, and he did not hesitate. "Huh?" he riposted, shuffling the cards.

She smiled, and both men shivered slightly. "Well, I do not understand it myself. But I have discovered that for some reason, many men find the customs of my planet extremely stimulating."

"What customs, my dear?" asked the old Colonel, clearly in better control than Fleming, though not by much.

"Well," she said demurely, "on Do It—my home planet—we have a society based on total sexual freedom. The theory is that if we can eliminate absolutely every sexual inhibition, we'll achieve a truly happy society."

Fleming put down the cards, got outside a couple of ounces of bourbon with considerable alacrity and dialed another, mashing down savagely on the button marked *Triple*. "Does it work?" he croaked.

Her smile disappeared, and he hastily searched his assets for something that might bring it back again. "Well, we do have one little problem. One of the first inhibitions to go was the incest taboo, and there weren't an awful lot of us to begin with. Daddy keeps saying something about our gene-pool being too small—anyway, we started having a lot of babies that weren't . . . quite right, one way or another." She frowned, then smiled again, and Fleming turned his triple into a single with one gulp. "But we figured out the solution, and that's when we instituted the custom that I was talking about.

"You see, Do It law requires any and all females to become impregnated by any off-worlder who offers them half a chance."

There came a sudden clatter as of dueling castanets, but it was only the sound of ice rattling furiously in two glasses at once.

The girl rose, traveled to the bar by a method that "walking" does not even begin to describe, and seated herself on the empty stool between Fleming and the Colonel, dialing a sombrero with a blood-red fingernail. Fleming essayed a gay smile and produced a simpering grimace; the Colonel tried to clear his throat, and plainly failed.

"Why don't you gentlemen come and visit me in stateroom 4-C tonight, say about 2300? Perhaps I can show you more of the customs of my planet." She turned to Fleming and added softly, ". . . and I may have a business proposition for you, young man. If you are interested . . ."

Fleming allowed as how he might be interested, and finished the remains of his drink. She rose, smiled at them both, and left.

Fleming and the Colonel looked at each other. As one they turned back to the bar and dialed fresh triples, bourbon and stengah. Raising their glasses in silent toast, they drank deep, then smashed the glasses against the bulkhead.

"2300 hours, she said?" asked Fleming at last.

The Colonel looked pained. "I say, old boy, I mean . . . dash it, both of us? Together?"

"How do we know what her customs are?" Fleming reasoned. "It may be necessary. Or something."

"Yes, but . . ."

"You want to blow it?"

The Colonel closed his mouth, opened it, then closed it again.

"Come on. Let's get back to your poker lesson."

Curiously, Colonel Enderby-Thwaite had a run of beginner's luck after that. At the close of the lesson, Fleming was a little startled to realize that he was down by about thirty credits, and what with one thing and another, he had much to preoccupy him as he made his way above-decks to the passenger level. But when his head cleared the hatchway of C-Deck and he saw a large, ferocious-looking Greenie tiptoeing down the corridor away from him with a blaster in its fist, he came instantly alert.

Greenies—natives of Sirius II—were the first and so far only alien race ever encountered by man; and the history of that encounter was not a happy one. Captured Greenies had been used as domestic animals for years before it was learned that they possessed intelligence, and even then it had taken a war with Sirius and several determined slave-revolts before men learned to see the green humanoids as equals. Even now, a hundred years later, many Greenies were still a little surly about it, and college students like Fleming had learned not to mess with them. This Greenie was large even for his race, and he was armed in the bargain.

But Fleming was not an uncourageous lad—there

was, in truth, a streak of romanticism in him that yearned for glory and danger, battle and sudden death. He silently eased himself the rest of the way through the hatch and began shadowing the Greenie.

From behind, a Greenie resembled nothing so much as the fabled, perhaps mythical Incredible Hulk, said to have walked the face of Old Terra centuries ago in the Age of Marvels—that is, roughly human, if one used Hercules as a comparison. From the front, Fleming knew, its humanoid look would be somewhat modified by the long, gleaming fangs and trifurcate nose, but it was otherwise remarkably similar to a human. One of the problems that Greenies had faced in fighting for equality was that they turned out to be crossfertile with human beings—and the males had enormous genitals.

This one wore native Syrian garb, shorts, and a fringed doublet, with an armband around its right bicep. As he padded silently behind the alien, Fleming noted uneasily that the armband would have been too big around to serve him as a belt. He hoped this Greenie was not one of those thionite-addicts. They were said to be violent and unpredictable.

The giant creature stopped before a stateroom door, and Fleming hastily ducked into an alcove. It placed an ear against the door and listened. Then it stepped back, brought up its blaster, and burned the lock off the door, leaping quickly through the smoldering doorway as it burst open. Fleming scurried down the corridor to the doorway, but stopped outside.

"Did you think you'd be allowed to keep all this money to yourself, Carmody?" he heard the Greenie boom within. "That's pretty selfish of you."

"You'll never get away with this," a human voice responded.

"You think not? You think perhaps you have friends on board? If so, they will be taken care of, Carmody.

This is the end of the road for you." The human voice rose in a shriek, there was a harsh, metallic *zzzzzap!* and then silence.

Fleming waited, paralyzed, in the corridor. From within the stateroom came the sounds of drawers being torn open, closets being ransacked. At last there was a triumphant exclamation, followed by a rattling noise.

Fleming remained hidden behind the opened door, frozen with fear. It was too late now to think of flight—the Greenie had found whatever it was looking for and would exit at any moment. He cursed his curiosity.

The alien stepped out into the hallway and stopped, separated from Fleming only by three inches of bulkhead door. It paused there a moment, and Fleming's heart yammered mindlessly in his chest. Then it strode off down the corridor in the other direction without closing the door behind it.

Fleming realized of a sudden that he had not breathed for some time, and debated soberly whether he ought to resume. He tried to move, discovered that he could, shrugged his broad shoulders and inhaled deeply. It cleared his head somewhat; he stepped round the bulky pressure-door and entered the room that the fearsome Greenie had left.

A stocky, balding man lay on his back on the floor of the room, an expression of agonized despair frozen across his features. His tunic was of extremely expensive cut and fabric, and his outflung hands were uncalloused and well-manicured. He did not appear to be breathing.

Fleming slowly crossed the room, bent down and reached for the man's wrist, intending to take his pulse. He recoiled at the touch and stood up. Carmody's wrist was quite cold. *Omigod*, Fleming thought, *omigod, what do I do now?* He was suddenly overcome with terror at the realization that the Greenie

prevented him from taking on the Greenie there and then. Fleming understood Art.

For all that, Nandi appeared exceedingly skeptical throughout Fleming's tale. "What would the Captain do if we got word to him?" she asked when he was finished. "Would he start a panic, perhaps endanger his passengers by trying to arrest the demon?"

"Of course not," said Fleming, who favored this alternative himself. "He could call the Patrol and have them send a cruiser to match speed and course with *This Train*. Let them capture the Greenie; they'll have sleepy gas and hypnodrene and vibes by the case. And in the meantime, we concentrate on lulling the Greenie into a false sense of security by preserving an air of normalcy."

"Why not just wait until we ground on Forced Landing and have it picked up as it debarks?"

"No good. We don't reach Forced Landing for at least seventy-two hours. Sometime between now and then, it may remember that it left that damned airtight door open. Even if nobody happened to glance in, sooner or later a meteorite-drill would make it pretty conspicuous indeed."

"Well, there goes your air of normalcy."

"Maybe; maybe not. That Greenie may be wasted on thionite, not thinking clearly. They often are. If we move fast . . ."

"Have you known many?" Nandi asked softly.

"Eh? Anyway, one way or another we've got to get word to the Captain before it decides to clean up after itself."

"I suppose you're right," the girl said grudgingly. She tossed cascades of lush brown hair casually back over one white shoulder and puffed a joint into life. "All right then. First, where is the Greenie now?"

Fleming had been waiting patiently for this line for five minutes. Precisely as Humphrey Bogart would have done it, he deadpanned, "Ten feet behind you," and *rolled* a joint of his own.

Her eyebrows rose quite satisfactorily, and if the orbs below them twinkled, Fleming failed to notice.

"How then shall we communicate with the Captain without tipping it off?" she asked. "I don't even know how one gets to see the Captain. Are you certain we've got one? My travel agent was a trifle vague."

Still Humphrey Bogart, Fleming essayed a humorless grin, producing a hideous grimace. "Relax. It's a snap. I've already taken care of it."

"You have?" she asked with new respect. "How?"

"Wrote a message on one of my rolling papers while I've been talking. I leave it in my plate, and the steward passes it up the chain of command to the Captain." In the ancient and bloody wars that had accompanied the birth of commercial space travel, the powermen's union had fared much better than that of the cooks. While cheap machinery was good enough to feed the passengers and crew, a human crew-member would feed the Converter with the leftovers, as well as the day's output of trash, performing valet duties in between to earn his keep.

Nandi's eyes widened, the increased candlepower melting the fillings in Fleming's teeth. "What a brilliant idea, Mister . . . what is your name?"

"Ayniss, Fleming Ayniss. My friends call me Flem."

"Listen, Phlegm, what do you suppose actually happened to Colonel . . . Benderby? Engleby?"

Fleming's dead pan acquired rigor mortis. "Enderby-Thwaite," he mumbled. "I don't know." He looked grimly across the room at the Greenie. "But that damned thionite-head was late for dinner."

"No, no. I mean, how do you know that the Colonel simply hasn't been taken with indigestion?"

"I knocked on his door on the way here," explained Fleming, stung that she thought he was jumping to conclusions like some romantic adolescent.

"Perhaps he has diarrhea, then, and ignored you. Or . . . or suppose he's in conference with the Captain right now? Let's . . ."

"Let's go to his room and wait for him," said Fleming, fighting to retain control of the situation.

They rose and left together, brushing past the Greenie with utter aplomb. Behind them, on the table, gravy began dripping lethargically across a cigarette paper half-buried in mashed potatoes, that read, "THIEF ON BOARD. CAPTAIN MUST KNOW. WASTE NO TIME."

Although the pair waited vigilantly in a lounge across from the Colonel's stateroom, Enderby-Thwaite had not returned by the time the ship's computer darkened the corridor lights for evening. Fleming and Nandi sat silent and motionless in the reduced light for ten seconds, then spoke simultaneously.

"My place or yours?"

Both blushed, but the phenomenon looked much more natural on Fleming. To his credit, however, his gaze never trembled, and if his knees did somewhat, that seemed only natural. A man's knees were supposed to tremble around girls like Nandi.

They ended up in her stateroom by Hobson's Choice—his was a mess. Hers was considerably neater; only the bedclothes were rumpled.

Nandi flicked on the light and crossed the room to the bed, sliding a trunk from beneath it. "You'll find some pot on the dresser behind you," she said over her shoulder as she attacked the clasps of the trunk, the part of her nearest Fleming describing a graceful figure-eight.

Fleming came back from a far place. "Er, no thanks."

"Oh, go ahead," she giggled. "It has to be all smoked up before we reach Forced Landing anyway. It's illegal there, remember? Go ahead and light up while I change into something more comfortable."

Eyes bulging with the sight of what she considered comfortable, Fleming turned obediently and began puttering with an elaborate water pipe. When he turned back, she was just stepping, out of the blue dress, humming ethereally. The narghile slipped from his nerveless fingers and landed on the floor with a crash.

She looked up; dimpled. "Oh, I hope I haven't upset you. It's only that I have nowhere else to change. Do you mind?"

"Not . . . not at all," croaked Fleming. A grapefruit seemed to have become lodged in his larynx somehow, and he strove mightily to swallow it. "G-go right ahead."

"You're so understanding." She beamed, slipping gracefully into what Fleming instantly realized was the most comfortable-looking garment he had ever seen. Intangible as a promise, its surface rippled with changing colors, flesh being the predominant tone. Wax began running out of his ears. "There, now, that's better." She lowered her gaze, drew in her breath suddenly. "Why, Fleming. I've . . . I've aroused you, haven't I?"

"Well . . . yes. I mean, yes, you have . . . uh, yes," he stammered.

"Oh, Fleming, how flattering." She grinned. "Do you know what I'm going to do to you?" She paused, looked thoughtful. "That reminds me, Fleming, I have a small favor to ask of you."

Fleming indicated a willingness to fetch a comet barehanded.

"No, nothing like that. I want you to keep something for me. My jewels." She returned to the bed, bent over the open trunk again (kicking Fleming's adrenals into overdrive), and removed a large package about the size of a shoebox. Opening it, she spilled fire onto the bedspread: several dozen gems that blazed with an unquenchable inner brilliance.

"Why, those are Carezza fire-diamonds," gasped Fleming, who had thought himself already as awestruck as possible. That many fire-diamonds would suffice to buy a fair-sized planet; one of them would have purchased *This Train* with enough change to pay for the balance of Fleming's education.

"Yes, my brave one. The hope of my planet. I have been sent to convert them into credits for the planetary coffers of Do It, so that we can begin a massive galaxy-wide advertising campaign to encourage immigration. The gene-pool, you know. Will you take care of them for me, until this inhuman thief has been disposed of?"

Fleming stood on one leg, opened his mouth, and made a gargling sound.

"I knew I could count on you." Nandi bubbled. "Lock them away somewhere, as tight as the Fist of Venus."

"The Fist of Venus?" asked Fleming weakly.

"You don't know the Fist of Venus? A standard accomplishment among my people. Among the women, that is. Here, let me show you."

She swept the bedful of diamonds to the floor, let her negligee join them. The floor became a riot of pulsing color. Smiling, she beckoned.

Fleming actually paused for a second. "If we were going to do this all along, why did you go through that business of changing into something more comfortable?"

"I thought you might enjoy the show." Nandi giggled. "Was I wrong?"

Fleming demonstrated that she had not been wrong.

Morning brought no word from the Captain, no sign of Colonel Enderby-Thwaite, and no steward at breakfast. To Fleming, who had begun the day with no sleep, it seemed that a definite negative trend had been established. In less than twenty-four hours he had become involved in at least one and possibly three murders, had taken on the responsibility of guarding more wealth than he could comprehend, and had learned that most astonishing and disappointing of truths: that there is such a thing as an overdose of pleasure.

The sandy-haired youth had annihilated six eggs, half a pound of home-fries, and two quarts of coffee before he felt reasonably safe in attempting rational thought. Now he rather regretted the undertaking.

It seemed obvious that the Greenie knew Fleming had witnessed its crime—the disappearance of the only two men to whom he had imparted his secret had to be more than coincidence. But how had it found out, in spite of all his circumspection? Fleming buried his head in his hands, and the answer smacked him in the face: his iridescent yellow boots, reportedly all the rage among the collegiate set on Secundus, gleamed up at him with a brilliance that was matched by no other footwear on the ship. At once Fleming remembered that the stateroom doors stopped four inches short of the deck, for a tighter airseal. The Greenie could scarcely have overlooked Fleming's toes—the mystery was that it hadn't murdered him there and then.

Well, that was that. Time to break cover and get to the Captain *fast*. Fleming had no idea where the Captain was to be found at this hour—his travel agent had been as vague about *This Train* as Nandi's—but he seemed to recall that anything above C Deck was "officers' country." *This Train* had been a luxury liner before she was a freighter, before all but one of her passenger decks were ripped out for maximum cargo space, and she bore quarters for a far larger crew than an automated tramp needed or could support. Considerations of mass distribution made converting that cubic footage into cargo room impractical. Consequently, finding the Captain could take on some of the salient aspects of finding the proverbial football in the asteroid belt.

"Unless he's actually in the Control Room," Fleming mused aloud, putting down his eighth cup of coffee.

"Beg pardon?" said Nandi, who had been absorbing

considerable fuel herself. "Who's in the Control Room?"

"The Captain, I hope," Fleming replied, then looked frantically round for the Greenie. It was not in sight. "Or one of the other officers," he finished in a whisper.

"Well, as I understand it, there's only one other officer *up* there," Nandi whispered back, "the Executive Officer."

"My God," gasped Fleming. "You mean they're the whole crew?"

"Well, there's the Chief Engineer, but I think he stays below with his converters and things. And the steward, of course, but we don't know where he is. The rest of the crew goes *clank* when you kick it."

"Well, how many passengers are there?"

"Aside from us, the Colonel and the Greenie are the only ones I've seen. The murdered man makes five."

Fleming had been counting on considerably more allies. He briefly considered stealing the Greenie's gun and blowing his own brains out with it. Being a hero was incredibly hard work.

But there was no help for it—no turning back. Resolutely he stood up, drawing Nandi to her feet. "Let's go," he said tersely, "before the Greenie shows up for breakfast."

They left, began climbing for officer country. Fleming paused when they reached A-Deck, frowning. "Look," he said, "the Captain or the Exec could be in any one of a couple-dozen staterooms—but either of them *might* be up in the Control Room. Why don't you pop up and check while I start searching here? That way we may save some time."

Nandi nodded. "All right, but be careful, my hero."

"My sentiments exactly."

He had tried about nine rooms unsuccessfully before it occurred to him that Nandi was a long time returning. Either she had found one of the officers, or . . . he sped back to the stairshaft, swarmed up three levels to the Control Room, and burst through the hatchway in classic unarmed-combat stance, ready to deal sudden death in any direction.

A mustached, competent-looking man in ship's uniform blinked amiably at him from one of the pilot's couches. "Sorry," he apologized, "I don't dance." He produced a green, odd-shaped bottle: three chambers hooked in parallel to a common spout. "Prepared to offer you a shot of Triple Ripple, though."

Fleming shook his head.

"Sure? Great stuff. Can't let the ingredients mix until you're ready to swallow, but when they do . . . oh, boy. I'm the Executive Officer, by the way. Name's Exton." He put out his hands.

"Where's Nandi?"

"Never heard of it; must be Capella way. Check the astrogational computer."

"No, dammit. Nandi, your female passenger. Hasn't she been here?"

"Nobody been here, no women for *damn* sure. Nandi? . . . don't believe I recall the lady."

"Then you've never met her," Fleming said positively. "Never mind, the important thing is that she's in deadly danger. Where's the Captain?"

"Aw now, the Captain wouldn't hurt no *passenger*. He's a gentleman."

Fleming gritted his teeth, then counted to ten and told Exton the whole story. The Exec listened attentively, tugging alternately at the Triple Ripple and his mustache. When the youth had finished, he leaned back on the acceleration couch and slapped his thigh. "Old son, that's the craziest story I ever heard. No wonder you don't want any Triple Ripple—it'd just bring you down. Let me tell you one about my Uncle Jed—true story too."

"Dammit, Exton, I'm telling you the truth. We've got to find Nandi before it's too late—and we've got to have the Captain flash the Patrol, so they can send a cruiser to rendezvous with us." He broke off, distracted by a sudden, indescribable sensation in his loins.

"Well now," drawled the Exec, "Captain Cavendish is something of an independent gent. Take a lot to make him call in the Patrol."

"You've lost sixty percent of your passengers and twenty-five percent of your crew," Fleming barked, tugging inconspicuously at his crotch. "What do you think the Captain would consider serious?"

The Exec blinked, looked thoughtful. "Well, your story certainly deserves checking, young fellow. Let's go below."

"Now you're talking."

"Reckon we'd best go to passenger country first and start checking staterooms."

"No," Fleming said decisively. "This Greenie is smart. We should go all the way down to the Converter Room and work up through the holds to the passenger deck."

"Sounds like a whole lot of work," Exton demurred.

"Listen, dammit, this Greenie snatched Nandi somewhere between B-Deck and here. It's obviously mobile and clever. The only way to nail it is to start at the bottom and work upward until we flush him out. If we let him get behind us we're finished." There was a peculiar look on Exton's face, but Fleming was too bemused by the drawing feeling in his groin to notice.

"All right," the Exec said reluctantly, "we search below." He rose, loosening his blaster in its holster. The sight of it reassured Fleming considerably.

"Whatever you do," he said as they left the Control Room, "look natural. We mustn't make the Greenie suspicious if we can help it."

"Okay," said the Exec agreeably, and began to sing a duet with himself.

Seeing Fleming's astonishment, he broke off. "Forgot, you don't know. I was born on Harmony, a pleasant little place where we feel that music is the bedrock of true culture. Most of us have had biomod work done on our larynxes, sort of improved on nature. I've got a five-octave range myself, and I can handle up to three voices at once. Handy—gets the women. And I guess it's how come I'm so partial to Triple Ripple, now I think about it."

Fleming puzzled over this as they made their way below. He had heard of biomodification even on the rather provincial planet of his birth, but he had always considered it a rather plasphemous attempt to distort the Creator's intentions. Now, however, he admitted to himself that there were advantages to more versatile vocal cords, at least. Exton was pretty good *a cappella*.

When they reached the Converter Room, the Chief Engineer was nowhere to be found. "Probably sound asleep somewhere, if I know Reilly," chuckled the Exec, but Fleming was filled with dark suspicion. They searched the Power Room thoroughly and found nothing.

"Well," said Fleming at last, "I guess that makes it fifty percent of the crew gone."

"Oh, listen here, young fellow. Reilly's around somewhere. He's got a lonely job, probably off brewing himself some rocket juice someplace or other." They started up the ladder.

"Listen, Exton," Fleming insisted as they reached the cargo level, "I don't think you're taking this whole thing seriously enough. There's a *killer* on board."

"That remains to be proved, son."

"But by the nature of the problem it may be almost impossible to prove before we're dead. *Won't* you call the Patrol?"

"With no evidence to show the Captain? Hah! I'll take my chances with this killer of yours."

"Well, keep your gun handy," grumbled Fleming, disgusted with the Exec's refusal to behave by adventure-story standards. Exton snorted, but drew his blaster. Together they began to search the cargo hold, a huge steel cavern piled high with stacked crates and tarpaulin-covered machinery of all sizes and shapes. The lighting was dim, and Fleming imagined crazed Greenies in every pool of shadow, but none materialized. Neither did Reilly. Finally every cranny had been poked into, unsuccessfully, and Exton started to return to the stairshaft.

"Wait," said Fleming suddenly. "I've got an idea. It seems to me that if I wanted to hide on a ship like this, I'd stay right here in the hold."

"But we've looked . . ." Exton began wearily.

"Not in the cargo itself," Fleming broke in. "I can see five crates from here that are large enough to fit the two of us in."

"Now hold on, young fellow," Exton protested. "I'm not about to start breaking open crates of merchandise that ain't mine to look for a killer I'm not sure exists."

"Well, I'm sure," snapped Fleming. Turning to the nearest crate of sufficient size, he slapped its pressure seals. Exton yelped in protest, but it was too late—the top of the crate slid open.

Fleming levered himself up on his elbows, peered down into the crate. "No luck with this one. Full of some kind of white powder."

"There, you see?" said Exton, wiping sweat from his forehead. "Sugar or something."

"No, wait!" gasped Fleming, excitement in his voice. "This is no sugar, Exton—it's thionite! Kilos of it."

"No crap?" the Exec said weakly.

"Sure. That lemony odor is unmistakable. Well, I'll be damned. There's more here than meets the eye. *Now* will you call the Patrol?"

"I reckon I'll have to do *something*," said Exton, looking grim.

Very suddenly a dark form detached itself from the shadows, landed on Exton's shoulders and knocked him sprawling to the deck. A gun-butt rose and fell, and Exton gave three cries simultaneously and lay still.

As Fleming dropped to his feet, numb with terror, the attacker rose and covered him with a vicious-looking little handgun. It tempered the relief with which Fleming noted that his assailant was human. "Who . . . who are you?" he stammered.

"Chief Engineer," snarled the other, "as if you didn't know."

"Listen, Reilly, give it up. You'll never get away with this."

"Shaddap and come here. You're going to carry this sleeping beauty right up topside to the lifeboats, and then the two of you are going for a nice little ride. Without an astrogational computer."

Fleming went cold. This sort of thing happened all the time in the adventure stories, but the hero was always prepared with something: a special plan, an unsuspected ally, a concealed weapon. Fleming had none of these; it had never occurred to him that the Greenie might have human confederates. The fact disgusted him.

Under the unarguable direction of Reilly's gun-barrel, Fleming heaved Exton awkwardly over one shoulder and began climbing. When they reached C-Deck he set the Exec down as gently as he could and began dragging him past the passengers' cabins to the lifeboat locks, noticing with numb indifference that two of the six locks were empty. "Out of shape," he gasped, and Reilly sneered.

When he had dumped Exton's limp and, by now, dusty form inside the first lifeboat in line, number three, he turned to face Reilly, who stood just outside the airlock with his gun leveled at Fleming's midsection.

"Can't we talk this over?" he asked. Reilly smiled, tightened his finger on the firing stud.

Suddenly voices came from behind him, and the Chief Engineer froze. "Hold it right there." "Patrol, put 'em up." "Drop it, Reilly."

At the last voice, Reilly suddenly unfroze again, and

his grin returned. "Nice try, Exton, but it won't work. That fancy throat of yours makes you a better ventriloquist than a Denebian Where-Is-It, but you can't fool me. If the Patrol really was behind me, they'd be calling me by my real name—which is *not* Reilly."

Exton sat up, shrugged. "Can't blame a fellow for trying."

"Maybe not," said Reilly, "but I can kill you for it." He broke off as the sound of shod feet on deck-plates came from behind him. "Say, that's pretty good. I didn't know you could imitate sounds too."

"He can't," said Nandi as she brought an oxy-bottle down hard across Reilly's skull.

The Greenie, Nandi explained, had spotted her on her way up to the Control Room and taken a shot at her that barely missed. Fortunately, she was able to elude the monster and hide in number four lifeboat, where she had remained in terror until the noise of Fleming dragging Exton into the neighboring boat had drawn her out.

"You've been very brave, Nandi," Fleming said approvingly as he checked the clip on Reilly's gun. "Well, Exton, now do you believe me?"

"Guess I sort of have to," the Exec drawled. "Wasn't for the young lady here, we'd be trying to astrogate through deep space by eye about now."

"What I don't understand," Fleming mused, "is why the Greenie took Reilly into cahoots with it. There's something going on here I don't understand. Well, anyway, it's past time we notified the Patrol."

"Suppose you're right," Exton agreed.

"Where do we find the Captain?" Nandi asked. "He should send the message."

"He usually hangs out on A-Deck," Exton decided. "We'll probably find him in the crew's lounge there, playing whist with the computer."

"Okay," said Fleming decisively, tightening his grip on his gun. "Let's go."

The three ascended together cautiously, Fleming in the lead, Exton covering their rear. As they climbed, they conversed in whispers.

"How do you suppose all that thionite you boys found ties in with Carmody's murder?"

"I don't know, Nandi. There are more questions than answers in this case." Fleming was a little short-tempered; the peculiar not-quite-pain in his groin was still troubling him.

"Maybe Reilly, Carmody, and the Greenie were in partnership on the thionite," suggested Exton from beneath them.

"Could be," Fleming agreed, pausing to peer cautiously over the hatch-coaming before exposing himself. "Then Carmody tried to doublecross them somehow—the Greenie said something about him trying to keep all that money to himself." He clambered through the hatch and reached down to help Nandi up, looking around for the Greenie.

"But where are all the bodies?" Nandi asked. "You found nothing in the Converter Room or the hold, and it would make no sense to hide them where there are more people around."

"I dunno, maybe the Greenie spaced them all. What you think, Exton?"

No answer.

"Exton!" Fleming stuck his head down through the hatch and looked around. The Exec was nowhere in sight.

Nandi gasped, began to tremble. Fleming set his jaw grimly and closed the hatch, dogging it as tightly as he could. "Let's go," he rapped, and began climbing again, pulling Nandi after him.

They found the Captain just where Exton had guessed they might, in a lounge on the uppermost of the two crew-levels, engrossed in a card game with a relatively simple-minded recreational computer. He was a patriarchal figure, massive and heavily bearded, authority obvious in both the set of his broad shoulders and the disrupter that hung at his hip. He rose as the two entered, bowed to Nandi, and raised a shaggy eyebrow at Fleming. "Yes? What can I do for you?"

"You can call the Galactic Patrol," said Fleming, and without preamble launched into his third retelling of the past day's events. The sincerity in his voice was unmistakable, and when he finished the Captain had a dark look on his craggy face.

"Your story is easily checked, young man," he rumbled. He closed a switch on the wall beside him and said, "Exton. Reilly. Report on the double to A-Deck rec lounge. Hop." His voice seemed to echo in the distance, and Fleming realized he had cut in the command intercom.

They waited for a minute or two with no result. Then the Captain rose to his feet and put his right hand to the butt of his disrupter. His mouth was a tight line and there was thunder in his eye. With his other hand he removed a remote control unit from his tunic, dialed a frequency, and said clearly, "Computer: broadcast, this frequency. 'Emergency. Emergency. Interstellar Vessel *This Train*, Captain Cavendish speaking. Killers on board, crew captured or killed. Request Patrol lock onto this carrier and rendezvous at once, repeat at once, prepared for armed resistance. Cavendish out.' Repeat and maintain carrier."

Fleming breathed a sigh of relief. Whatever happened now, the Patrol would be here soon. He began to relax—then stiffened as he realized that the intercom was still on. The Greenie must have heard every word! He waved frantically to get Captain Cavendish's attention and pointed to the wall-switch. Cavendish paled, put down the computer-relay link, and slapped the switch open, but the damage was done.

"Look," rapped the Captain, "I've got to get down to the lifeboats and make sure that damned creature doesn't escape before the Patrol arrives. You stay here, barricade the door, and don't stick your head out until I knock shave-and-a-haircut. You've got to keep this young lady safe," he put in as Fleming began to object. "I'll be all right—I've had some experience with hijackers before."

Fleming reluctantly agreed. He hated to lose out on potential heroics, but protecting the fair maiden was definitely a duty no hero could dodge. Besides, it was safer.

As soon as the Captain had left, Nandi came into his arms and captured his lips in an urgent, demanding kiss. "Hold me, Fleming," she breathed, "I'm so afraid."

"Don't worry, Nandi," Fleming reassured her with all the bravado he had left. "I'll keep you safe." His arms tightened protectively around her.

"Of course you will, my hero," Nandi said. "Just as you are keeping my jewels safe. Where did you put them, by the way? I've been meaning to ask you since that Reilly almost . . . I mean . . ."

"I understand," he said quickly. "I should have thought of that. They're in my stateroom, under my pillow. Who'd ever . . ." He trailed off. Nandi's hands had begun to wander, and while that confoundingly indefinable sensation had not left his groin, Fleming discovered that whatever it was did not interfere with performance. *What the hell,* he was telling himself, when suddenly something caught his eye and made him go rigid from the waist up as well. "My God," he breathed.

"What is it?" asked Nandi, sensing that he was no longer responding.

"That communications relay the Captain used. It's not set for emergency band at all—way off, as far as I can tell. Cavendish must have accidentally dialed the wrong frequency. Good Lord, if I hadn't noticed . . ." He let go of Nandi, picked up the device and reset it, then repeated the Captain's message as best he could remember. "That was too damned close. We would have waited for the Patrol till we were old and gray." He frowned. "I'd better let the Captain know what happened, so he doesn't do something foolish if the Patrol is late in showing up."

He activated the intercom. "Captain Cavendish?" No reply. "Captain, this is Fleming. Come in, please, it's urgent." There was no response at all for a long minute, and then the speaker came alive.

"Mr. Ayniss," came the unmistakable booming voice of the Greenie, "I would advise you not to meddle in criminal matters. They don't concern you."

Fleming jumped a foot in the air, his pulse-rate tripling instantly. Somehow the giant killer had gotten the drop on Cavendish, turned the tables again. The youth made a quick decision, a decision based on pure heroism.

"Wait here," he barked, killing the intercom again, "and don't let *anyone* in. I'm going to do what the Captain failed to do—keep that damned creature here until the authorities arrive. We can't let it get away."

Nandi began to protest, but Fleming ignored her and stepped out into the corridor, gun in hand. He was genuinely terrified, but a cold anger sustained him and steadied his weapon in his inexperienced grasp. He felt partially responsible for the carnage that had resulted from his discovery of the original murder, and he meant to avenge the crew and passengers of *This Train*. The Colonel, the steward, the Exec, the Captain, Nandi, all had been innocent victims, ordinary decent folks attacked without knowing why, given no chance to defend themselves. The murdering alien would pay for its crimes—Fleming intended to see to it. He made his way to the lifeboat locks, his peripheral vision straining to meet itself behind his head.

Unfortunately, it failed in this endeavor. As he approached the lifeboat locks, agony exploded in the back of his skull and extinguished the corridor lights one by one. He never felt the deck smack him in the face.

The blow had been startling, but he was considerably more surprised to regain consciousness, alive and unharmed, his gun still nearby where he had dropped it. He reclaimed it, rose shakily to his feet and staggered to the locks.

All six lifeboats were gone.

Got away, dammit, thought Fleming. *Probably fired off all the other boats to make itself harder to track.* He was furious, with himself as much as with the Greenie. His only consolation was that the murderer had been careless enough to fail to finish him off. He decided to make sure Nandi was all right, and headed back up to the lounge.

Nandi was not all right. At least, she was missing from the lounge. Fleming knew one timeless moment of pure fury, the frustrated rage of undeniable failure. The Greenie had obviously taken her along as a hostage in case the Patrol caught up with it.

The youth sank down into a chair and buried his head in his hands. He was bitterly sorry that he had ever heard the word *adventure*, and he cursed the nosy curiosity that had precipitated this slaughter.

After a long, black time he began to think again. Numbly, he decided to go below and check whether the Greenie had removed the thionite from the hold. Perhaps it had been in too much of a hurry.

But when he reached the hold, he heard noises from close at hand and melted quietly into the shadows, his gun growing out of his fist.

It was the Greenie, it had to be! How, Fleming couldn't imagine and didn't care; the song of blood rushing in his temples had a one-word libretto: vengeance. He smiled grimly to himself and clenched his gun tightly, peering with infinite caution around the fender of a halftrack farming vehicle.

The Greenie was just resetting the seals on the opened crate of thionite, an ominous expression on its face. Fleming took careful aim at the massive head, but before he could fire, the alien strode rapidly to the stairshaft and climbed above. Fleming slipped from concealment and followed it, reaching the shaft in time to see the Greenie step off two levels above, on C-Deck.

Narrowing his eyes, Fleming ascended noiselessly to C-Deck, just quickly enough to spot the killer entering Carmody's stateroom, the scene of the original murder. He waited, hidden by the hatch cover, until the creature had exited and turned a corner. Fleming padded silently after it. As he passed Carmody's room, he glanced in, and was not even mildly surprised to discover that the corpse was missing. It figured. The Greenie was housecleaning.

Fleming intended to do a little housecleaning of his own.

He eased around the corner with care, but the corridor was deserted. The nameplate on the third door he came to read, "Rax Ch'loom, Sirius II." Jackpot!

Fleming took hold of the doorlatch, paused for a long moment to bid good-bye to his adolescence, then yanked open the door. The first thing he saw was the Greenie, surprised in the act of changing clothes, literally caught with its pants down. The second thing he saw was Carmody's body on the bed, neatly trussed up with nylon cord. A part of him wondered why the Greenie would tie up a corpse, but the majority of him simply didn't give a damn.

"This is for Nandi, you bastard," he said clearly, and aimed for the trifurcate nose.

And then something struck him between the shoulder-blades, smashing him to the deck. His chin hit hard enough to drive a wedge of black ice up into his brain, where it melted, turning everything to inky dark.

* * *

"Crap," Fleming said as consciousness returned.

"You bet, old son," said a pleasant baritone. "Several fans-full of the stuff, in fact."

Fleming looked up, startled. A smiling lieutenant of the Galactic Patrol knelt over him, smelling salts in one hand and a vortex disrupter in the other.

"Did you get him?" Fleming cried. "Did you get the Greenie?"

"Ch'loom? Hell no, Mr. Ayniss, but we got damned near everybody else. God-damnedest thing I ever saw—a freighter torching along practically empty, and six lifeboats full of crooks heading away from it in different directions like the Big Bang all over again. We picked 'em all up okay, but what I'd like to know is what put the wind up all of them? Their stories don't make much sense when you put them all together."

Fleming shook his head confusedly, allowed the Patrolman to help him to his feet. "I don't understand," he said weakly. "Lifeboats full of crooks?"

"Sure," said the Patrolman, holstering his sidearm. "First one in line was a Colonel Underwear-Waist or some such, claimed you were the first sucker in ten standard years to catch him stacking the cards."

"Huh?" gasped Fleming, thunderstruck.

"Yep. Old-time card-sharp, according to our computer records. Been working the tramps for years, ever since the regular lines got on to him. How'd you tumble to him, Ayniss?"

"Uh," Fleming explained. He tried to recall the exact wording of the message he had slipped into "Captain Galaxy Meets His Match" a hundred years ago. "Who was next?"

"Next was the ship's steward, chap named Blog. Says you found out he was rifling staterooms and threatened to tip off the Captain, so he lit out as soon as he could. We found a lot of boodle with him—guess you've got a reward or two coming. Then there was an engineer who claimed his name was Reilly, but he turned out to be a guy named Foster, wanted for murder over on Armageddon. Had his fingerprints changed, of course, but he couldn't afford biomod work on his retinas. According to him, he heard you and the Executive Officer talking, realized you were on to him and stuck the two of you up. Then, he claims, somebody else sapped him, and he woke up alone in number three lifeboat, which he did not hesitate to use.

"But the next customer was the Exec himself, Exton is it? And under questioning he broke down and admitted smuggling thionite on board to sell at Forced Landing. We found the thionite just where he said it would be. Say, did you know he's got a modified voice-box? Cursed you out in three-part harmony.

"But the strangest of the bunch was Captain Cavendish himself. He was really surprised to see us—kept insisting that he'd called us himself and he was *sure* he'd used the wrong frequency, which doesn't seem to make much sense. But he was so flabbergasted he slipped up and mentioned what frequency he *had* used. Just for fun we broadcast, 'All clear, come ahead,' on that frequency, and a whole gang of pirates walked into the surprise of their life. Apparently Cavendish was in cahoots with them on some kind of insurance fraud scheme, figured to let them rob the ship without a fight. We've got 'em all, and we didn't lose a man."

"What about Nandi?" Fleming asked groggily. "She has to be honest—she gave me a fortune in jewels for safekeeping."

"Nandi *Tyson*—'honest'? Say, we've been looking for her for years, ever since she started passing out counterfeit Carezza fire-diamonds in the outworlds. She was the last one we picked up—she had those diamonds with her, by the way—and boy, was she ever mad at you."

Fleming's head spun. "Does this mean that the diamonds are worthless?" he asked.

The Patrol lieutenant had studied classical humor in college, but even as the phrase "Ayniss and Nandi" exploded hilariously in his brain, he felt a flash of compassion for the crestfallen youth and kept a straight face.

"Put it this way, Ayniss," he said gravely. "Yes."

"But—but what about the Greenie? Didn't you get him too? He's the one that started all this madness."

"No, my friend," came a booming voice from the doorway, "I am afraid you did that all by yourself."

Fleming whirled. The Greenie stood there smiling, a gun at its hip, a Patrol officer at its side. "Get it," Fleming screamed at the lieutenant, "it's a murderer."

"Ch'loom a murderer?" the officer said dubiously. "That's a little hard to believe."

"I tell you I saw it," gibbered Fleming. "The damned thing killed a man named Carmody."

The Greenie's smile deepened, exposing more fang. It stepped aside, to reveal Carmody standing behind it, demonstrably alive. Their wrists were handcuffed together. Fleming's mouth opened, and stayed that way.

"Allow me to introduce Rax Ch'loom, Official Equalizer of Carson's World," said the lieutenant. "My name's Hornsby, by the way, pleased to meet you."

"Equalizer?" mumbled Fleming dazedly.

"Sure," Hornsby replied cheerfully. "Rax showed up on Carson's World about thirty years back and commenced stealing from the rich and giving to the poor. The idea caught on so well they institutionalized him—gave him legal immunity from prosecution, quasigovernmental status, subsidies, the works."

"What did the rich do?" exploded Fleming.

"Squawked like hell," Rax grinned. "There wasn't much else they could do."

"But don't the rich hold the political power?" asked Fleming, stunned.

"Hey," Rax replied, "we got democracy. *Lots* more poor people than rich people on Carson's World."

"Sounds like a crazy place to me," Fleming snapped, his confusion turning to unreasonable irritation.

"Oh, I dunno," Hornsby intervened. "You get hungry, you go see Rax. You start hogging, Rax rips you off. Sounds pretty comfortable to me."

"Suppose you rob a rich man, and for want of capital he's utterly ruined the next day?"

"You get hungry, you go see Rax."

"But there's more to life than food."

"Hey, listen, Rax don't steal no women . . ."

"Other things."

"Like what?"

"Carmody here thought he could take thirty million credits out of circulation," Rax boomed contentedly. "Not a chance." Carmody snarled impotently.

"But I felt his wrist," Fleming objected feebly. "It was cold."

". . . as a corpse's wouldn't have been for at least an hour," Rax pointed out. "I put him in a cryonic stasis for my own convenience, and spent the whole rest of the voyage trying to figure out what in the name of the seven bloody devils of Old Terra you were doing."

Fleming gave up, began shaking his head. "Then it's all over?" he asked resignedly.

"Er . . . not quite, Ayniss," Hornsby said with curious reluctance. "There's one more little matter. Did you and the Tyson woman . . . ? I mean, did she . . . ? Did you . . . ?"

"Well, yes," Fleming admitted, remembering that he did have at least one thing to be proud of. "She's from Do It, you know."

"You knew that and still let her?" gasped Hornsby, his jaw dropping.

"Hell, yes. She said Do It was based on total sexual freedom, so as to eliminate tension and frustration. It sounded like a good idea to me."

"It sounds like a good idea to me, too, but that's not

what Do It is like. It's a world full of fanatical feminists, not hedonists. All the women have had the same biomod work performed on them."

"What kind?" Fleming asked, feeling that strange and indescribable feeling in his crotch again.

"Uh . . . well, you may as well know. It has to do with modifying the ovum, giving it the mobility and the seeking instinct of a sperm cell, with some of the parasitic characteristics of a tapeworm. Only a psychotic female-supremacist could have conceived of it." He broke off, embarrassed.

"Well?" said Fleming. "*Tell* me, dammit."

"I'm afraid, Mr. Ayniss," said the Greenie with genuine compassion, "that you are pregnant."

TIDBIT: afterword to "When No Man Pursueth"

I'm sitting in the New York apartment of Ben Bova, then editor of *Analog* magazine, drinking his scotch. "What made you decide to move to Nova Scotia?" he asks. So I tell him.

I was eighty percent decided when I left that province, after spending perhaps a week there visiting friends. I found the Annapolis Valley a pastoral paradise, spattered with beauty and inhabited by decent, honest people. My friends had no way to lock their home, had never considered doing so. We went shopping in splendid old Annapolis Royal, and when my pal Charlie had more groceries than he could conveniently carry, he set them on the fender of his truck and kept going. As I followed, bemused, I noticed that the keys were in the ignition. Our next stop was the drugstore, where the pharmacist squinted at Charlie's prescription and said that he could sell him the brand name specified for five dollars or the identical substance in an off-brand for three dollars, what was Charlie's pleasure? He followed us out into the street when he discovered that he'd accidentally short-changed us.

I was one mind-blown New Yorker.

So I got off the return ferry in Bar Harbor, Maine, with the intention of inquiring about emigration when I got back to New York. Write a letter to the Canadian Embassy or something . . .

Then I boarded the Greyhound bus.

All went well until I changed to the New York express bus in Boston. A woman with three children and no brains boarded the same bus. So did two barefoot thirteen-year-old habitués of the bus terminal, who had seen her flash an enormous wad of bills there. She bedded down her kids and went to sleep herself, protecting her purse by placing it on the floor under her seat. The youths copped it, of course.

Now, this deed was observed (we all learned later) by one of three white male college freshmen, on his way to the bathroom. He gave no sign, but once in the toilet he scrawled, "CRIMINALS IN REAR OF BUS. MAY BE ARMED. INFORM POLICE," on page 27 of the *Mad* magazine he was carrying. He returned to his seat, urging his friends to read the terrific article on page 27. They yelped and were frantically shushed. Together they put the torn-out page into a tin of chocolate chip cookies. After a time one of them yawned, said loudly, "I think I'll see if the bus driver would like some chocolate chip cookies," and wandered up front. The bus driver read the chocolate chip cookies, and his eyes widened. He nodded grimly and winked, and at the next toll booth he passed the note, tin and all, into the startled hands of the toll-booth attendant ("Oh," Ben said to me at this point in the anecdote, pouring more scotch, "I get it. Toll-house cookies"), who subsequently called the cops, who called the state cops, who called the Highway Patrol, and you might not believe this but the message got a little garbled in transmission. The Highway Patrol received the impression that the bus was held by armed hijackers with machine guns. A laborious trap was prepared; instructions were given to all the attendants at the next toll booth down the line.

I knew nothing of this, you understand—only six people on the bus knew that anything was going on. I woke from the brain-flensing half-sleep of bus travel to the realization that the coach was stopped by the side of the road. My seatmate was a charming sixty-eight-year-old Swiss/German lady of enormous education and wisdom, with whom I had had a fascinating discussion earlier. She was an experienced world traveler, who said that she had visited every major city in the world without ever witnessing anything like the squalor and filth of parts of New York City. I had been attempting to defend my native land against this and other criticism, and the effort had put me to sleep. I asked her now what was amiss, and she told me that the bus driver had debarked with a fistful of paper towels, muttering something about an oil leak. I thanked her and started to go back to sleep.

A state trooper literally leaped up the steps of the bus, landed on both feet with a loud crash, aimed a loaded, cocked and unsafetied hogleg down the aisle, and roared, "FREEZE!"

Everyone in the bus promptly leaped three feet in the air and came down dancing.

The little old Swiss/German lady suddenly saw jackboots, uniform, and gun, instantly flashed *Gestapo!*—and she had a number tattooed on her forearm.

She lost all her English, clutched my arm and ear, and began to gibber in three other languages.

The trooper saw his mistake and changed tactics. "All right," he bellowed, "everyone off the bus. Now! *Move!*"

A stampede ensued, in which he barely retained the cocked pistol. Kids were getting trampled. The little old lady was having a fit. I was beginning to get pissed off.

As fifty of us fell out the door, we were confronted by other armed cops, some with riot guns and sharpshooter rifles, and they pointed to six nearby black-and-whites and screamed, "Everybody down behind those cars!" and we all flashed *sniper?* and dove headlong for cover (all this really happened, I swear to God) and then there was this really *long* pause. I was trying to calm the Swiss lady, assuring her that this was *not* the Germany of her youth and those jack-booted troopers were *not* Nazis. About that time they began pulling passengers out from behind the cars for questioning one at a time, and we could hardly help noticing that they were only interrogating black people. "You are *sure*?" she asked me. I had no reply.

Sadly, I can't fault their technique on pragmatic grounds: the third and fourth blacks in line were the two kids, who could not satisfactorily explain why they had six hundred dollars in twenties and no shoes (and, by the way, no weapons. They had never had any, except in the mind of a media-child freshman who knew perfectly well that all those black kids carry knives). They were marched away in handcuffs, the rest of us were ordered back aboard the bus and sent on our way with no explanation, and on the way to New York we got the story from the college kids.

And when we docked in Port Authority terminal, I stashed my suitcase in a coin locker and *ran* uptown to Canadian Manpower and Immigration for an application, filled it out and submitted it on the spot. (They turned me down, too, but that's another story.)

Ben opened up another bottle of scotch. "You know," he said thoughtfully, "there's the makings of a damn good science fiction story in that yarn."

"I don't see it," I said argumentatively. "Bus into spaceship doesn't translate—and besides, the story's essentially a downer." What I was really saying was, "Write a little *more* of my story for me, Ben."

He obliged. "The story's a downer because you've got a bus load of innocent victims traumatized. But if you had a spaceliner full of thieves, thimbleriggers and con men . . . and the same intrepid hero went around hiding cryptic notes in *Mad* magazines . . ."

"And the one guy he thinks *is* a crook is the only honest citizen in the bunch—" I said excitedly. "Ben, that's an *amazing* idea."

"No," he corrected smugly. "That's an *Astounding* idea."*

And so, pausing only to kill the bottle, I staggered home and wrote "When No Man Pursueth."

But that's not why I love Ben Bova.

He let me name the hero Fleming Ayniss—that's why.

*For those of you who don't know (both of you), the original title of *Analog* magazine was *Astounding*. And one of its competitors is a magazine called *Amazing*. Ben's sense of humor is nearly as primitive as my own.

5
NOBODY LIKES TO BE LONELY

The room looked quite comfortable when they brought McGinny in and left him alone. He had seen pictures, and knew what it was. But in his guts he could not believe that it was a cell.

It didn't look like a cell. It didn't taste like a cell, or feel like one, but most of all it didn't look like one. McGinny had been in jail once before, in this same county, and the cell then had borne all the classic hallmarks: bars, mildewed concrete walls, barred windows, an absurdly large lock, and miserably inadequate sanitary provisions consisting of a seatless toilet which stubbornly refused to flush and a badly cracked sink which exuded brown, rusty water.

But then, that had been so long ago that the charge for which McGinny had done time was possession of marijuana. That statute, while it still existed, had not been enforced in over ten years.

And in the meantime, prisons had changed. They had had to, of course. The Attica Uprising and the Tombs Rebellion, the Joliet Massacre and the Battle of New Alcatraz had been unmistakable signs that the traditional approach to penology was obsolete. A criminal population approaching thirty percent of the total simply could not be herded together and kept safely subjugated without the very sort of brutalization which an informed public would no longer tolerate.

But what if they were not herded together?

So it was that the room which met McGinny's eyes

now was in appearance a pleasant, modestly appointed studio apartment—with a few anomalies. The convict seated himself in a remarkably comfortable, high-backed psuedo-leather armchair, padded with God alone knew what, and surveyed the unit which would be his universe until the time-lock on the room's only door ran out, ten years from now. *Lookit all the cubic,* he told himself wonderingly. Maybe this wouldn't be too bad after all.

The time-lock itself, not unnaturally, was the first thing that held his eye. It was set just below the apparently open window which was cut into the door of his cell. All that faced on his side of the door was an inverted triangular plate with rounded corners, small horizontal grooved slots in each corner. The overall effect was damnably like a skull.

"Pleased to meet ya," McGinny told it, returning its sour grin.

The window above the plate measured about three by three, and appeared empty of glass. So did the window on the opposite wall behind McGinny, but both were in fact enclosed with a synthetic material (trade-named "Nothing") which was so transparent as to appear invisible. It could not break, crack or get dirty. The second window looked out on a small courtyard, pleasantly landscaped with ferns and lush grasses, bordered by three fifteen-story wings just like the one which held McGinny's cell. The seven hundred and fifty windows of each were opaque, and McGinny knew that his, too, seemed opaque from the outside. He sighed.

To his left was a bed, consisting of a mattress on top of a sealed box-spring which was clamped to the floor. Although the room's climate-control system made bedclothes superfluous, the penologists had been thoughtful enough to realize that a man (or woman) felt better with something over him as he slept. Hence they proved a sheet—made of paper. Above the bed were two horizontal slits, each about a half-meter wide. The upper one would dispense either paper sheets or paper clothes. It was activated by placing a used sheet or garment in the lower slot, which led to an incinerator somewhere in the bowels of the prison. Two pillows lay on the bed, each a featureless sponge.

Filling the space between the head of the bed and the corner of the room was a closet without a door. It had no transversing pole from which to suspend hangers, nor did it have hangers. Instead, suits of paper clothing—there were four of them—hung from small extrusions of plasteel high on the rear wall of the closet.

In the opposite corner, behind McGinny and to the right, was a spacious desk with voicewriter and drawing pencils. Above the desk was a reader which would display any book requested, page by page, so long as it was stored in the prison's central computer. Much of the fiction available was speculative, the authorities having decided that it would be all right to allow prisoners *some* form of escape. (McGinny knew that lately, the majority of science-fiction writers were ex-criminals, some of whose output was quite disturbing. Or perhaps that was not a new development.)

To the left of the desk was a quadio console, also computer-supplied, its four speakers represented by darker areas at four corners of the ceiling. Available tapes ranged from classical through rock to flash, with side trips into gregorian and neojazz. The console was nearly featureless: one spoke one's choice and selected tone and volume with simple slide switches. In appearance, therefore, the console resembled a washing machine with two small horns.

Directly adjacent to the quadio was the Automat: an equally large cube, with a serving platform let into its front and small slots on either side which dispensed rubber cutlery. It too was voice-activated, and was fed through the floor from a master unit which supplied the Automat with raw materials. Save for the absence of a slot into which to deposit one's quarters, it was identical to the Automats to be found on the average street corner—from McGinny's angle of vision at least. From the other end of the room one could have seen the unmistakable, time-honored shape of a toilet bowl, let into the Automat's left side. It drained to the prison's basement where paper and waste were filtered out and the remainder routed to the master food unit. This saved the taxpayers millions of dollars annually.

McGinny snorted, ceased his inventory of the room and rose from his chair. He went to the small sink on the right of the cell door and regarded himself in its "mirror," a glassless reflective surface. As McGinny was one of many who had elected to inhibit his beard, there was no shaving unit next to the mirror; his hair would simply have to grow for the next ten years, or until he became sick enough to warrant the cutting open of the time-lock to permit a doctor to attend him. The doctor played a lot of golf.

Familiar, coarse features stared back at McGinny, restoring his confidence. His head was large, with a cap of wiry brown curls resting on elongated ears. His eyes were set close against a blunt nose, and his over-full lower lip gave him a pouting, petulant expression. As he saw again the room whose reflection surrounded his own, the pout became almost a sneer. These were the most spacious and luxurious quarters he had ever inhabited—few in the overcrowded world of 2007 had it so good.

Ten years? he thought, cheerfully. *I'll do it standing on my head. Elbow room, privacy, food cooked for me . . .* He frowned. *Sure will miss beer, though. And the fems.* His contentment beginning to fade, he returned to the armchair and dropped heavily into it. He found his gaze fixed on the window set in the cell door. It was strange—the window on the opposite wall looked out on open space, this one onto a plasteel corridor. And yet the exterior window gave a view of a false freedom, sculpted to make McGinny and other thousands feel better. In the corridor, men walked. Somehow, freedom was that way.

He shifted, scratched his crotch and considered the quadio. It seemed to him that his first choice in this cell was a significant event, demanding contemplation. He imagined himself ten years hence, narrating his prison saga to an enraptured fem with eyes like saucers, saying, "And do you know what the first thing I played in that taken place was?" This'd better be good; he'd hate to have to lie to her.

After a time he addressed the quadio. The room filled with the sound of a frenzied 4/4 piano solo from Leon Russell's legendary last album, *Live At Luna City*. Bass and moog came in together as the Master of Space and Time hurled his anthem:

"I'm just tryin' to stay 'live—and keep mah sideburns too."

Legs trembling, vaguely enjoying the play of cool air across his sweat-sheened, slender body, Solomon Orechal lay in the utter relaxation called afterglow and surveyed his bedroom. In so doing, he also surveyed his dining room, his living room, his kitchen, and his car—all at the same time.

He sighed, for perhaps the dozenth time that day; just as, in fact, he had sighed with an almost rhythmic regularity on every day since he had first moved into his own Mome, from the comparative spaciousness of his parents' fish-and-see apartment. As the popular name indicated, a good efficiency was hard to find these days, but the Orechal ancestral apt (the building dated all the way back to 1957) had been in family possession since before the Housing Riots—as the axe-scar and single bullet-hole in the door attested. Solomon had told himself often in the last two years that he had been a fool to strike out on his own. But the lure of adventure and the challenge of living wherever he could find a parking space had been enough to pry him from the four-and-a-half-room home of this youth.

Besides, it was awkward, bringing your girlfriends into the bathroom to be alone.

Apropos of which:

It's very strange, thought Solomon. *I know just what she's going to say now . . .*

"Sol, why can't we do the Truth Dope?"

. . . and yet there's nothing déjà vu about it.

Beside him on the narrow bed, Barbara raised on one elbow, half-leaned across him. Sleepily, earnestly, she brushed the hair out of his eyes and repeated, "Why won't you do Truth with me, lover?"

. . . even down to the soft but oh so insistent tone of voice, the way she lets her left breast brush me; and it's just nothing at all like déjà vu . . .

She was still talking, and there was that in her voice which acts directly on the glands, but he was miles ahead of her, his attention two levels removed, contemplating the frustration of Moebius's Band with what seemed a poignant bitterness. Vaguely, he monitored the persuasions and importunities, dropping a grunt here and there and looking impassive, until he heard the line he had been patiently waiting for.

". . . how," she was saying, timidly and inevitably, "can I help but think you're afraid of the Truth?"

His timing was magnificent.

"Afraid of the truth?" he asked quietly, paused. "What we just did . . . wasn't that the truth?" He brushed his fingertips along the underside of her belly, and she shivered. "Are you suggesting that that wasn't real? That we were just *fucking*? Because it sure seemed to *me* that we were making love. Maybe I was wrong."

He had her now, he knew it from the look on her face, but somehow he couldn't summon up the old elation, the sense of triumph. Mechanically, he moved for the coup-de-grace: now that you've stirred up the emotions, throw in a little pseudologic and you're home free.

"You know why I don't do Truth Dope, man. I've told you a dozen times. I'm not afraid of the truth, I'm afraid of the *dope*."

She made one last try. "But Sol . . ."

"Now don't start, Barb. We've been through this, kark it. There's a mountain of evidence for each side, just like there always is when a new drug comes out. The law says it plays hob with your motivations, and the heads say it clears your vision. The law says it rots your body, and the heads say it's a lie. You know what happened with pot." (It hadn't been until 1986 that it was proven that marijuana could cause tuberculosis. No real problem, as they had TB licked by that time— one shot at 12 and you couldn't get it if you tried—but it was too late for an awful lot of smokers who had thought that all the evidence was in by 1975.) "I lost my mother to TB, and I plan for the rest of my life to take the conservative opinion wherever possible. No thanks, Barb. I'll take my Truth the sloppy, human way, through inference and deduction. Maybe I'll be wrong a lot more often . . . but maybe I'll have a lot more often to be wrong in.

"Besides, I don't need any proof that you love me— even though you're trying to get me to do something I don't think is safe, to reassure *you*. Things like just happened here a couple of minutes ago are all the 'proof' I need."

There was, of course, nothing she could say to that, and she even apologized, but somehow even as he mounted her to prove again the depth of his love by the strength of his hips he knew that the subject was not closed, and that someday she would back him into a corner he couldn't talk his way out of, and on that day they would share the drug that made dissembly impossible, and she would leave him, just like all the others.

He moaned, but she misunderstood and held him tighter.

McGinny tried for the fifth time to cut the leathery soyburger with his rubber fork. This time the disposable plate danced on the serving platform and he nearly lost the meal entirely. He swore a hideous oath and flung the fork angrily from him, but with the blind malignance that inanimate objects display when a man is in a towering rage, it bounced from the plasteel wall and dropped with an absurdly loud, high splash into the toilet.

He rose quickly, cursing with a steady, monotonous rhythm. *Taken stuff tastes enough like rubber already,* he thought savagely, plunging his thick hand into the bowl. He was just too late to save the fork; the cell's designers had reasoned that a flushing mechanism could fail—a serious calamity in a time-locked room—and so that bowl simply emptied itself constantly, at a gentle speed which McGinny had not quite beaten.

Swearing louder now, he straightened and walked to the sink to wash his hands. He could not for the life of him understand why he felt that the water there would be any cleaner than that which laved the bowl, and it irritated him immensely.

Of course he burned his hands. But by that time the anger had reached the point from which one either tremblingly descends, or begins throwing things. He had few things to throw, and none he could spare. He counted to ten, then chanted Om Mani Padme Hum, and gradually the black rage subsided, at least to the point where he could see through the red haze.

Make the karkin' silverware rubber so we can't snuff ourselves, he thought, *and look how much good it does. I'm really filled with the joy of livin' now.*

Finally he walked back to the automat, sat down in the desk chair which stood before it, and picked the soyburger from the plate on the serving platform.

It was cold.

"GodDAMNit," he exploded. "Sunnabitchin' machine s'posta keep the taken stuff hot, just my *fuckin'* luck to get the one don't work for TEN TAKEN YEARS!"

There was nothing for it; the soyburger was all he would get until tomorrow morning. Growling, he raised it to his mouth and ripped off a piece with his teeth.

"Hi, there."

He whirled, his hand absurdly cocking the soyburger like a weapon. There in the window of the cell door, above the skull-like time-lock, was a face. A person!

McGinny ran to the door, flinging the soyburger into a corner. "Hello!" he shouted, and then pulled to a halt before the door, suddenly embarrassed. They looked at each other for a while, McGinny seeing the young kid, maybe twenty, with long blond hair and a Fu Manchu mustache, *looks like one o' them Trippies, oh Jesus, I hope he likes to talk.*

"What are you in for?"

"Embezzlement," McGinny said automatically, a million questions that he could not form coherently enough to ask buzzing in his brain.

"Oh," said the youth, adjusting a uniform cap on his shaggy head. He seemed somehow just slightly disappointed. "I guess that must be pretty interesting stuff, embezzlement. I get to talk to all kinds of interesting people on this job. Once . . . once I talked to a rapist."

He almost seemed to be licking his lips, but McGinny was beyond noticing. He managed to stammer, "Hey, look, buddy . . . what . . . I mean, who are you? What are you doing here? How often do you come around? What . . . hey, how come I can even hear you in here?"

The kid chuckled. "They've got a two-way sound system on the door, man. Didn't you know? Listen, don't freak, I'm like, the guard. You didn't think they'd leave you alone with nobody to check on you, did you? Suppose you conked?"

"But," McGinny said, "I mean, do you come around a lot? *Can you stay awhile and talk?*"

"Oh, sure," the kid assured him. "That's why I took this job, man. I'm into people, where they're at, like. All I have to do is walk around and talk to interesting people, and I only gotta cover fifteen guys a day. See, if you want to know the truth, the job's welfare."

McGinny understood. The work-and-wage system as a means of distributing wealth was on its last legs— there simply wasn't enough work to go around, and the population continued to climb. As a last-ditch stopgap, the government had taken to making up idiot work so that there would be sufficient jobs available to keep the traditional economic system staggering on, but the farce was becoming more obvious every year. What more obvious example than this young Trippie, "guarding" men in sealed plasteel cells to earn his living.

But at this particular moment McGinny was overwhelmingly grateful for the continued sham. It was accidentally providing him with the means of maintaining his sanity.

"Listen," he said urgently, "listen, kid, if you'll come around and talk to me a lot, I'll . . ." He paused, baffled. He had nothing to offer. "I'll be grateful," he finished lamely, desperate with fear that he would be rejected.

"Sure, man," the kid grinned. "I like to talk. Mostly I like to listen. I'm interested in the criminal mind and all. I'll bet you've got some interesting stories to tell."

"Yeah, you bet, kid. I got the most *goddamned* interesting stories you ever heard in your life!" He paused again, embarrassed by his fervor.

"Hey, listen, man," the kid said softly. "I know how it is. Nobody likes to be lonely."

And he smiled.

* * *

The mome ahead completed its business and gunned away noisily, and Sol pulled his own vehicle smoothly up alongside the Chase World Bank. Rolling down the forward driver's-side window, he addressed it.

"Solomon Orechal, 4763987IMHS967403888.453, license NY-45-83-299T."

The Bank, which bore a remarkable resemblance to a vacuum cleaner making love to a garbage can, asked San Francisco a question, received a reply, and answered without a millisecond's hesitation, "Sir?"

"Request additions and alterations allotment, three thousand dollars and zero cents; travel allotment, five hundred dollars and zero cents."

"Purpose?"

"A and A: Fortrex cooling unit. Travel: To Lesser Yuma."

"Justification?"

"Profession: entertainer."

"Type and Credit Number, please," the Bank said a bit more respectfully. Its voice was like a contralto kazoo.

"Folksinger. Number SWM-44557F, ASCAP. I'm my own agent."

This time the machine actually paused. Barbara squirmed on the seat next to Solomon, twisting her hair nervously. "Aren't you going to get it, lover?"

"Relax," he said easily. "The Bank's got to consult a human for this. Judgment decision required. It's bound to take a minute or so; they've got to decide if I'm worth shipping across the country."

"Oh, Sol . . ."

"Now don't worry, Barb, I told you. If the Bank says no, I'll use my own credit and we'll go just the same. Now relax, will you?"

The squat machine spoke up. "So ordered," it said emotionlessly, "and good luck to you, sir. Have a pleasant time in Lesser Yuma."

"You got it," she said excitedly as Solomon engaged gears and roared away from the Bank, "oh, baby, you got it! When can we go?"

"Get centered, mama," he answered as he slid the huge mobile home smoothly into the freeway traffic. "There's a lot of things we have to do first. We've got to get the cooling unit installed, gotta cop a big block of food, got to get the engine overhauled and tuned. Gotta say good-bye to our parents. It'll be a couple of days, easy. Less if we bust ass."

Behind his practical words Solomon was immensely pleased with himself. Barbara had been difficult lately, carefully avoiding any mention of Truth Dope but finding more and more reasons to sulk. But he'd managed to find something to distract her. She'd never been out of New York State in her life, and travel held a fascination for her, as for so many. A similar feeling had been responsible for Solomon's decision to buy a Mome in the first place, and so he was somewhat excited about the trip himself.

And, too, his ego writhed with gratification that his performing record was in fact impressive enough to make the Bank invest in his relocation to an area where performers were scarce. Consciously he had never doubted the outcome, and he would never admit his subconscious doubts, but it felt good to *know*.

You had to be good to be a performer; it was one of the most sought-after jobs in the country. It wasn't only the tremendous prestige, nor even the almost orgasmic egoboo that applause brought. It was simply that the first time you saw drab, apathetic faces come alive during your set, the first time you made some of those thousands of crowded, useless people a little more content with their lot, somehow you never again felt that gut-ache of uselessness quite so sharply yourself.

"Sol," Barbara said softly, breaking into high reverie, "do we have to start . . . right away?" Her soft fingers traced a question mark on his thigh.

"Mama," he mock-growled, "I'll never be. *that* busy!"

And no one was more surprised than he when, having found a place to park the Mome, he failed to achieve an erection.

"How did you ever come to be an embezzler, Mr. McGinny?"

"I embezzled."

"No, I mean why?"

"Because I wanted some money."

The kid was impervious to sarcasm. "What did you want the money for?" He adjusted the guard's cap that looked so incongruous atop his shaggy mane, his hand stroking his mustache on the way down in a mannerism which McGinny suspected he could learn to hate sometime in the next ten years. "I mean, it isn't like way back in the seventies when people were hungry."

"Listen, what is this, a quiz show or something? I mean, what's it to you?"

"Oh, I'm just curious, is all. I mean, there's nothing much else to do on this job but talk with you fellows. Anyway, crime interests me, you know? Like the things that made you end up . . . in here."

"Well, it's none of your taken business, how do you like that?" McGinny snapped. The kid made as if to turn away, and suddenly McGinny almost panicked. The kid was a pain in the joints, but he was better than nothing, better than the tangled, tormenting company of McGinny's own thoughts, of his self-recrimination and his frustrated rage.

"No, wait, kid. Listen, I'm sorry, please wait. You . . . you don't want to lift off so soon. C'mon, look: a guy gets a little hot under the collar sometimes, you ask him personal questions. I didn't mean any offense."

The kid half-turned back to the door, stroking his mustache again.

"Look, it was like this, see? I'm an accountant, I was, I mean, and they pulled an audit at the wrong time. No big story—I just got caught with my hand in the cookie jar. Could have happened to a dozen other accountants, just happened to be me, that's all."

"Why'd you have your hand in the cookie jar?"

"I needed the money." There was a pause, and the kid turned to walk away again.

McGinny cracked. "It was a fem, dammit."

The kid turned back again, smiling now. A gentle smile. "Yeah?"

McGinny gave in. Maybe the kid was right—it might help to talk about it, straighten his thoughts. In any case it was certainly better than trying to think of something new to play on the quadio. Or something to dictate into the voicewriter, which stubbornly refused to do anything more than repeat his own thoughts back to him.

"It was like this: I had to get my hands on a whole lot of money at once to shut this fem up. She had something on me that could have ruined me, had me by the hairs, and she loved every minute of it, the little slot. She had it in for me, but she needed green more than she needed my scalp, and she didn't even care if I got burned gettin' it. 'You're an accountant,' she says. 'You can get it.' Sure. Easy. Ten years easy, and she walks away, laughing. I had a chance, I'd be in here for murder right now."

The kid was all ears now, face almost pressed up against the cell window like a child at a candy-store window. "What'd she have on you?" he breathed.

McGinny turned bright red. The kid didn't bother to pretend to leave again; he simply waited. After a time the prisoner answered him.

"See, she was . . . she was pregnant without a license, and she was far enough along she was going to start showing any day, and she said when they hauled her in she was going to name me in the affadavit. The pregnancy fine alone could have ruined me, let alone the Lifetime Child Support without even a welfare option. I mean every man's entitled to welfare, isn't he? You can see what a jam I was in. I just had to have the green—she said if I gave her enough money to keep her and the kid until she could leave him with a sitter and go to work, she'd tell the Man she didn't know who the father was."

"I don't get it," the kid said cheerfully. "What was the sweat? You'd have beat the heat easy. Kark, they couldn't pin an Elsie's on you—it's your word against hers. Unless there was a photographic record of the conception . . ." his voice trailed off with just the faintest suggestion of a leer.

McGinny shrugged, made a face. "Well, maybe they couldn't have pinned an L.C.S. on me, if it came to that . . ." He seemed disinclined to continue.

"Then I don't understand why you took such a risk," the kid persisted.

"Well," McGinny said reluctantly, "I . . . I got a wife and kids."

"Oh," the kid said brightly. "Have you got a picture of them?"

"No I have not got a karkin' picture of them!"

"All right, all right, don't jump salty. I can take a hint. Sorry if I bothered you." The kid gave his mustache a final tug, turned, and walked out of view down the corridor. Suddenly terrified, not wanting to be alone with his memories, McGinny beat against the door with his fists.

"Wait, damn you, *wait*! Hold on a taken minute, I didn't mean to shout at you. Hey, listen, I'm sorry, wait, come back, please come back. Come back, you bastard you, don't leave me alone. You sonofabitch, I'll cut your heart out, *COME BACK!*"

Footsteps echoed faintly down the acoustically muffled hallway.

McGinny looked down at his hands stupidly. They ached terribly, and the heels of them glowed an angry red. He went to the mirror on shaky legs, tried a sickly grin, then whirled and threw himself across the bed, and very suddenly he was crying, the wild, racking sobs of a child.

Sol looked around at the hundreds of prairie rats who made up a cross-section of the population of this particular sector of Lesser Yuma, brushed the guitar strap out of the way of his wrist, and adjusted the microphone with a feeling of growing desperation. He wasn't reaching them, he just couldn't get it on for this audience, and he felt a frustration which was growing familiar of late.

It's the people, he told himself frantically, tuning up to stall for time. There was plenty of parking space left in the deserts, and hence a trouble-free existence for Mome-owners who could afford cooling gear. But the thousands who had flocked to the vast barren espanses had learned quickly that boredom was the price of exurban existence. They looked to entertainers like Solomon to keep them going, but the wary ennui they brought to a concert depressed him so much (he told himself now) that he just couldn't seem to get into his music tonight.

In desperation, he seized upon a song that summed up his mood precisely, one of his own. For the first time in his career he didn't care how the audience liked it, whether it was what they wanted to hear. He hurt, and so he sang.

> *This time next year . . .* *
> *I will have won or lost*
> *This time next year . . .*
> *my bridges all*
> *will be crossed*
> *I'll either be*
> *in headlines*

*Music and lyrics in Appendix A, p. 310

Or standin' in
the breadlines
It all depends
on how the dice are tossed

This time next year . . .
I will be up or down
Far away from here . . .
or still hung up in town
I'll either be in clover
Or barely turnin' over
It all depends on how
the deal goes down

I feel it comin' on—
it's O so close now
Wonder if it's
bad or good
Hope it isn't gonna be
an overdose now
Really wish I knew
where I stood

This time next year . . .
I'll either win or lose
This time I fear . . .
I'm on a short, short fuse
I'll either be together
Enjoyin' sunny weather
Or suckin' up
an awful lot of booze

He trailed off, fingers stinging from the harsh, emphatic runs. The catharsis of the blues left him literally exhausted, but the pain was reduced to an empty, fading ache.

The applause nearly frightened him out of his wits. From then on he had them, had them in the palm of his hand. Having made them cry, he could now make them laugh or clap or dance or anything he had a mind to. He had shown them that he shared something with them, and now they could empathize, let themselves be taken with him along whatever musical road he chose to pick.

It felt good.

It was on the way home, joyfully breaking the speed limit and humming snatches of his closing number, that he heard the news from Barbara.

"Sol?"

"Yeah, kitten? Here, have a toke."

"Later." She waved the joint away. "Sol, the clinic called while you were onstage. I came out to get my shawl and played back the message."

"Oh."

There was a pause.

"Sol, they said . . . the results were negative."

A longer pause, long enough for humiliation to turn to anger.

"Well, what the hell is that supposed to mean? Why, they're full of shit. Negative! What is that supposed to signify, it's all in my karkin' mind? Is that it?"

She was silent, and his fury boiled over.

"ANSWER ME, GODDAMMIT! Is it all in my mind?"

"Sol, I don't *know*, baby, I don't know. Maybe they made a mistake." She was crying, soundless tears highlighted by oncoming headlights, and he flung the joint out the window in disgust.

"Don't make excuses for me, you taken slot! It's no big deal. So the results were negative, so there's a little something I got to work out in my head is all. You know I've got it. I just have to get it back."

He drove on furiously, concentrating on the road until his eyes ached from squinting. They left the Mome colony behind, took a seemingly abandoned side road up into the hills. The road swerved treacherously beside sheer precipices at some points, but Sol had his control back now, and his hands on the steering wheel were unnaturally steady. The ponderous Mome was like a live thing under his hands, and he drove it with a grim determination. Eventually they passed through a great shadow-filled crevice between two walls of granite, and came out upon a ridge overlooking a great valley, invisible in the darkness.

There were only seven or eight Momes parked here, clustered around the natural mountain spring which surfaced in this unlikely spot. It was sufficiently long that there was at least an acre for each of them. Solomon had been lucky to find this place; the few who had tended to keep their mouths shut. *We are all very happy here,* he thought savagely, wheeling the huge Mome to its parking space.

He parked, shut down the engine, extruded the watersucker and threw power to the house generator. Pushing the button that dropped the seat-back flat, he got up and walked to the back of the Mome, flinging himself down on the bed without a word.

Barbara got up and walked slowly back to the bed, sat down on the carpeted floor beside it.

"Sol, what do we do now?"

"What the kark can I do?" he said, voice muffled by the pillow.

"Well, as far as I can see, there's only two things left. Analysis, or . . ."

"Or the Truth Dope," he snarled, lifting his head to throw her a venomous glance. "Get my head candled or my chromosomes scrambled, that's the choice, huh?"

"Well, all I know is I'm pretty karkin' sick and tired of masturbating," she shot back, and then gasped.

He winced.

"I'm sorry, baby," she said pitifully. "You know I didn't mean that."

"Well, it's true, and there's nothing I can do about

"Sure. Hey listen, wow, I meant to ask you. You never told me about how come you let that fem talk you into taking the green." The jailer tugged at his mustache and regarded McGinny expectantly.

McGinny turned, took a few steps from the window. Then he frowned and turned back resignedly. "It's like I told you: she was going to stick it to me."

"Yeah, but she couldn't prove a thing. Or could she?"

"She didn't have to prove it. I told you I got a wife and kids, didn't I? What do you think my wife'd do, I'm down in Paternity Court? What do you think my boss'd do? Bigshot Z.P.G. supporter, he'd toss me on the street in a minute. It ain't like if I sold illegal dope or run over somebody stoned. You can't get fired for criminal record anymore. But an unlicensed pregnancy? A third kid? Don't make me laugh. She didn't have to prove a thing to finish me off."

"Yeah, I guess I see . . ." said the kid. "But one thing I don't understand . . ."

"You don't understand nothing. You never been married. I'd have done anything to keep Alice from leaving me. Anything." His voice broke. "I . . . I loved her."

"That's what I don't understand," the kid said eagerly. "I mean, if you loved her so much, how come you topped this other fem? I mean, sure, everybody likes variety once in a while, but you must have a House in your neighborhood, you must have had the money."

"Hey, listen, I never paid for it in my life," McGinny said proudly. "I mean, half the thrill of love is in the conquest." He had read that somewhere.

"So then, since your wife was already 'conquered' she didn't turn you on?"

"Of course she turned me on. I told you I loved her, didn't I? But there was this fem I met at the Automat, worked in the same building, and she looked like she never had it, you know? So I called her up that night, invited her out for a drive."

"Top her that night?" the kid exclaimed.

"Well, sure," McGinny said modestly. "You know, I kind of always had good luck with virgins."

"Plural? You mean there were others?"

"Not too karkin' many others. I told you I loved my wife," McGinny said suspiciously.

"But you said . . ."

"I know what I karkin' said," McGinny barked.

"Okay, take it easy. I was just asking. 'Cause I thought you meant . . ."

"Well, keep your thoughts to yourself. Jesus, you ask a lot of dopey questions. What's the matter, you got nothing better . . ." His voice trailed off as he caught himself. "I mean, what makes you so taken curious?"

"Oh, I just wonder a lot. You know, how come you're in there and I'm out here and all—I've always been kind of *philosophical*, you know? Into people, like I said. I mean, we all start out the same, and some of us do things others don't. I guess I'm just curious about what makes people tick. How come she got pregnant?"

"Huh?"

"I mean, don't you use anything?"

"Well, sure, but I mean, I didn't know. Hell, first date and all, I . . . I just figured she'd be using something. Nice piece like that . . ."

"But you said she looked like a virgin."

"Well, that's it, see? How was I supposed to know she'd spread right off like that?"

"But you just *said* you always had good luck with . . ."

"Get off my case, will you? I'm telling you, this fem was a slot. She . . . she told me it was all right, see, because she wanted to get me by the pills, pump me for green, get it?"

"Look, I don't know, you were there and I wasn't,

but frankly that sounds like a load of used food to me," the kid said evenly. "You told me all she asked for was support until she could work again, didn't you? And just for that she was willing to take the rap and lose her own Welfare. Doesn't sound like a slot to me."

"Get out of here, you fuzz-faced stuffer! Who the hell asked for your opinion, anyway? Go on, get taken before I . . ."

"Before you what, bro?" the kid asked softly. "You can't get out of there, can you? You can't even snuff yourself to embarrass me. I'm not a captive audience, but you're sure a captive performer. I don't understand what you did, and you're going to explain it to me. Sooner or later."

"I'll see you in hell first," McGinny shouted, almost gibbering.

"Sooner or later," he repeated, tugging at his mustache.

McGinny's eyes widened, and he placed a hand on either side of the window. "You're enjoying this, aren't you? You little hark, you're really enjoying this!"

"Does that matter?" the kid asked softly. "Does it really make any difference whether I enjoy it or not? All I'm doing is asking you questions. The answers you already know yourself, right? Or you couldn't answer the questions. I'm not putting any words in your mouth—just asking questions so I can understand why you did what you did. All I want," he said simply, "is the truth."

"You want it, you clinical little bastard, but maybe *I don't*," McGinny snarled.

"Oh, well . . ." said the kid, shrugging. "There I can't help you, Mr. McGinny. I mean, even if I don't ask you another thing, you've got ten years to go, and there's no place to hide in there. How long you think you can duck the truth?"

"Forever, you lousy bastard," McGinny roared. Get

out of my life, go on, get the hell out of here." He turned away in dismissal, began pacing the room angrily. *I don't have to take this kind of sewage! I'll write to the Warden, to my Congressman, to* . . . he stopped suddenly, struck by the obvious. Prisoners lost all their civil rights—including access to the postal computer network. His voicewriter lacked the familiar "Transmit" key. There was no way for him to get a letter to *anyone*, unless the kid agreed to take it down for him and deliver it.

Somebody else has got to come by, sooner or later, he thought frantically. *A maintenance man, somebody!*

No one had so far.

He was trapped, pure and simple, trapped with this shaggy punk kid with his words that twisted the truth into lies and made you feel like you'd done something wrong, like you deserved all this instead of merely being caught up in a web of circumstances that could have happened to anybody. *The little stuffer'll be back, to pick at me and twist everything all up. Enjoys it, like he was pulling the wings off flies, like* . . .

He spun around angrily, and the kid was still there, his face framed in the window over the skull-like time-lock.

"Spying on me, you . . ." McGinny groped for words.

"No," the kid murmured. "Just . . . just *observing* you."

McGinny howled.

The drug which Solomon Orechal's age knew as Truth Dope had been known to man for hundreds of years before a single word was ever written about it. Known, that is, to some men.

The first words written about Truth Dope appeared in the middle Twentieth Century. Author William Burroughs passed on a legend of unknown origin concerning a forgotten tribe in the trackless wilds of South America who used a drug he called "yage," which induced temporary mental telepathy between its users. The brief mention was too preposterous to be taken very seriously, of course, but there were many in those times who took preposterous things seriously. Rumors traveled the junkie grapevine, apocrypha rode the dealers' trail, and the A-heads spoke in whispers of yage.

In vain. Yage existed, and its ridiculous Lost Tribe as well. But they were not exactly lost.

They were hidden.

For the telepathy that its users experienced under the influence of yage was more than the ability to send and receive messages without material aid. It was rather a total dissolution of all the walls surrounding human consciousness, a complete opening of minds one to the other, providing the first and only escape from the solitary confinement of the human skull. It was a melding of personalities, a stripping away of all cover.

Two people who took yage simply had no secrets from one another. At all.

Secret thoughts, inner motivations, hopes, shames, dreams, pretenses, likes and dislikes and the true inner feelings of that part of the heart whose name is unpronounceable, all were laid bare to a partner in the yage experience.

That the drug should have remained so perfect a secret for so many hundreds of years was not in the least surprising. Realizing what they possessed, and its potential for good and evil, its discoverers—the Kundalu—adopted a policy of isolationism utterly simple in execution: anyone they did not recognize was apprehended, and yage stuffed down his throat.

Then they either killed him or married him.

This delightfully uncomplicated system lasted until 1984. Inevitably, the Kundalu were discovered, by a real estate developer looking for a place to put 2,650 condominiums. Over twelve hundred years of self-knowledge on a level unknown to mankind at large had made the Kundalu wise and canny indeed—175 of the condominiums had been built and fifty-three sold before the clearing crews stumbled across the Kundalu village.

The strange and humble Indians would not leave the land where holy yage grew, nor permit its razing.

They resisted the developer's half-hearted attempt to learn their vestigial spoken language, lest the secret of its growth be somehow wrested from them. He, in turn, was impatient—and out there in the bush, no sanctions could be applied to him—he was, after all, building *dwelling units*. He slaughtered the simple Kundalu to the last man.

It chanced that four of the crew assigned to demolish the primitively beautiful village of the Kundalu were welfare clients—counterculture types who recognized the ceremonial bowls of dried leaves they found for what they were: a communal drug. The foreman found them inside a structure like a decapitated dome, open to the skies but closed to the gaze of passersby, and he understood enough of the joyous babbling he overheard to shoot all four of them dead.

In six months he and the developer had a small but established corporate identity in the underworld of big-time drug traffic. In a year, the developer had him killed. Within four years, the developer was outselling the quasilegal giant, Speed Inc., and was giving even the mammoth completely legal International Marijuana Harvesters a pain in the balance sheet, despite the fact that Truth (as yage was brand-named) was still on the Illegal List.

The usual controversy flared in the news media, freighted with a larger than usual bulk of ignorance, for very little indeed was known about Truth Dope. In time the substance might completely overturn many time-honored concepts of personal privacy, many institutions of law and justice, many truisms of human psychology—but at present absolutely all that was known about it was that it was curiously resistant to chemical analysis, and that no more than three people could safely share the drug. The stress of mingling identities with a larger number were severely unhinging; the ego tended to *get lost*, and the secret of finding it again had died with the Kundalu. Before that had been proven to the counterculture's cynical satisfaction, many communes ended in gibbering insanity.

Nor did many triads flourish. By its nature Truth became a couples' drug. Thus:

Solomon and Barbara sat naked in the rear of the Mome, facing each other in lotus. The windows were opaqued, the roof transparent; the mobile home was open to the skies but closed to the gaze of passersby.

"Should we smoke?"

Sol considered this at length, shrugged. "I don't see why not. The parts to be opened go deeper than pot can reach. Maybe it'll relax us. This is going to be a little scary."

Barbara caught his nervousness, mulled it over carefully. "Sol . . . you're really jumpy about this, aren't you?" A flash of insight: "You've done Truth before, haven't you?"

"Why ask? You'll know for yourself in a little while."

"Sol . . . Sol, maybe you're right. We don't have to rush into this. I don't . . ."

"You don't want to know?" Sol burst out. "After all the pleading and convincing you're scared of the Truth? Oh, no! Have a few tokes and then we'll get to it. I'm not going to call this off now, and then wait to see how long it is before you want to know again, before you start hinting and then urging and then demanding. No way, mama. We're doing Truth today."

Barbara lowered her eyes, and busied herself searching for the Grassmasters. She found a crumpled pack on the right-hand service shelf over the bed and passed them to him. Current social etiquette required the woman to wave the joint alight, but Solomon had chosen to smoke GMs specifically because they did not have ignotips, and had to be lit by hand. He enjoyed the archaic ritual of striking fire with his hands and placing it where it was needed, and spent a not insignificant portion of his income on the hard-to-find matches. Now more than ever, she sensed, he would want that feeling of control.

He accepted the marijuana impassively, producing a box of wooden matches from the pocket of the tunic which lay beside him on the bed. By his other side lay the ancient, handmade Gibson J-45 which was his comfort and sometimes his voice, and Solomon struck a match along the silk-and-steel A string with a quick snap of his wrist. Echoes of whispering giants overflowed the sounding-box, and Solomon sucked flame through the filtertip joint with a sharp urgency.

He passed the joint to Barbara, cupping it protectively in his hand. Reaching to take it, she was struck for the first time by how much in him was conservative, if not reactionary. His independent thinking had struck her until now only as an evidence of the creativity she admired and loved in him; all at once she realized how much of him yearned for an earlier age. He cupped the joint as if wary of detection—yet pot had been legalized long before his instincts were trained. He played an acoustic guitar in an electronic age—certainly it sounded mellower than contemporary instruments, but mostly it was *older*. In a dozen innocent mannerisms she detected for the first time an undercurrent of yearning for the uncomplicated past, when men still controlled their destiny. *If I keep pulling insights like this*, she thought, gulping smoke, *I won't need Truth.*

And it was true. Expecting imminent truth, her mind was revving up, extending the sensitivity threshold of its own built-in truth detectors, trying to approach both drug and experience as honestly and openly as possible.

She passed the joint back to Solomon, who took it impassively, emptying his lungs for a second hit. He would not meet her eyes.

She watched his bare chest fill as he drew on the smouldering cigarette, and became unaccountably aware of the weight of her own breasts. She looked down at them, and it was only when she observed that her nipples were swollen that she remembered that before the night was out, Solomon's impotence should be over at last. In a vivid flash of memory she saw again the look of his eyes when orgasm took him, and she shivered.

"Barb."

She looked up. He was holding out the joint, breath held tightly. Brushing hair from her eyes with a vague hand, she took the joint, which was burned down close to the filter.

She inhaled sharply.

Very suddenly, the air began to sparkle, and a gentle buzzing filled her head. "Whoops, I'm stoned," she said and giggled, taking another puff.

"Say, you must have been smoking some of that there merry-wanna," Solomon said gravely.

"Well, of course, ye damn fool," she crowed, spraying smoke. "How else would I get stoned?" They roared with laughter.

Sol retrieved the joint from her relaxing fingers and stubbed it out in an ashtray. Still giggling, he slid open a sliding panel in the wall, removed an Oriental figurine: a carven dragon with sparkling eyes. He touched it under one wing, and its mouth opened wide. Prisoned in its lower fangs was a blue capsule.

Solomon tilted the dragon. It spat the capsule onto his upturned palm.

Babara stopped giggling. "Oh," she said. "Yes."

Solomon met her eyes. "Yes."

He made a long arm, pulled open the refrigerator,

and removed a plastic flask, red with white logo. "Better take this with soda," he said judiciously. "Taken stuff tastes worse'n peyote."

He could have read that in a magazine, she thought.

He put the flask of coke on the bed between them, shifting his weight carefully to avoid spilling it. He dried his sweaty left hand on his thighs and broke the capsule open onto his palm. It made a powdery pile of gray veined with green, fine-grained and dry. He held out his hand.

Barbara reached, gingerly bisected the pile with her thumbnail, sweeping the two portions far apart. Looking up at him one last time, she bent close, licked one of the two doses from his hand, and grabbed for the coke. She made a face. "Oooooh!"

He nodded gently as she gulped coke, then took the flask from her. Eyes on the remaining powder, he licked and gulped coke in almost the same motion. When he had swallowed, he put down the flask, wiped his hands on the bedspread, and took her hands in his.

"Okay, mama," he said with great tenderness, suddenly vulnerable. "Here we go."

McGinny came howling out of sleep, flailing wildly with leaden arms.

"Goddam skull-faced kid," he shouted, and then fell back exhausted, drenched in sour sweat. Coherence came slowly to his thoughts, and he was torn by an unbearable craving for a cigarette. He tried to masturbate, and could not.

He rolled finally from the bed, padded to the bookviewer, and selected a book at random, falling heavily into the chair. He stared at the displayed title page for a few moments, reached out to punch for the next page, and slapped the set off instead. He buried his face in his hands and wept.

Nerves stretched wire-tight, he shook with racking sobs. He dug his knuckles into his eyes, but could not banish the haunting palpebral vision of Annie beside him on her bed, naked and vulnerable, cringing under his wrath (his baby planted now in her belly). He ground the heels of his hands against his ears, but could not banish the sound of her tears as she begged him for emotional support ("You *said* you were going to divorce her. Mack, I need you with me on this—it's *our baby*.") He beat at his skull with his clenched fists, but he could not deny the memory of his decision to "borrow" enough money from his company to leave town, to go underground, and leave the whole impossible tangle of his life behind.

And above all, he could not shut out the voice of the blond kid with the incongruous hat, could not seal the holes that soft voice blasted through McGinny's carefully-wrought fortress of rationalization. When the mind refuses to face truth, it very often knows what it is doing: a high truth-level is only tolerable to saints and those sinners who, loving themselves, have learned how to forgive themselves. But McGinny no longer had any choice.

For the kid never attacked in any overt way, never quite gave him a justification for his helpless rage. He just . . . asked questions, and McGinny could not keep the answers from leaping unbidden to his mind.

Nor could he forget them now. The jailer's soft voice, hideously amplified, seemed to fill the cell, as it had for days now.

"Well, I don't know, Mr. McGinny. You say that security and prestige were your goals, but doesn't it seem like you already had them both? And yet you weren't satisfied . . ."

"So then you're saying sex is kind of like a power trip for you, aren't you?"

"Well, why didn't your father divorce her then? I would have."

"Then Annie's probably having a pretty rough time of it now?"

"I get it. You were afraid to leave Alice . . . No?"

"But isn't that just a fancy way of saying . . . ?"

"But you just said . . ."

"But didn't you just . . . ?"

"But I thought . . ."

"But . . ."

McGinny burst from the chair with an animal howl and swept the desk clean of paper with clawed hands, swinging his arms wide and scattering sheets in all directions. "I'll kill you," he shrieked, and tore at his hair.

He lurched around the cell, kicking and punching at the unyielding fixtures, slamming his shoulder into the wall with whimpered oaths. He beat on the surface of the quadio, snapping off both controls, and the machine roared into life. Shorted somewhere within, it picked its own tape, at peak volume. The selection was old, stereophonic, activating the rear speakers only—it balanced perfectly. The ear-splitting voice of Leon Russell plaintively asked:

> *Are we really happy*
> *with this lonely*
> *game we play?*
> *Searching for*
> *words to say*
> *Searching but not finding*
> *understanding anyway*
> *We're lost*
> *in this masquerade.*

McGinny staggered, his hands over his ears. He could not shut out the song. He lay down on his back and smashed at the quadio with his bare heels, and it went dead with one last shriek.

As he lay panting on the floor, his ears still ringing, he opened his eyes to see the kid watching him from the door window.

McGinny began to sweat profusely. He struggled to

his feet and looked wildly around the room. *Rubber silverware, paper sheets, no razor, GO AWAY, KID!*

"Say, did I hear noise just now? Kinda late to play the quadio, isn't it, Mr. McGinny? Oh, I bet I know. You got to missing Alice and the kids, didn't you, Mr. McGinny?

"Hey, Mr. McGinny! What are you . . . *hey!*

"Oh, holy shit."

"Oh, wow."

The kid's face pressed closer.

The drug came on very slowly at first.

For what seemed like hours, Barbara felt only a gradual numbing of her extremities, a slow falling-off of communication with the nerves and muscles of her limbs. She and Solomon gazed deep into each other's eyes, motionless in lotus. She yearned to let her gaze travel downward over his body, but she maintained eye contact tenaciously, as though afraid of opening a circuit that was being built between them.

Very suddenly she was blind. Almost immediately, all tactile sensation vanished from her body. Adrift in crackling black, she could no longer see or touch anything in any direction. Although she had learned enough from friends and media reports to be expecting this, it still took her by surprise. She yelped.

As from a great distance, she heard Solomon's voice reassuring her, needlessly explaining that they were only experiencing a repression of distractions, that it was only a drug which would wear off, the standard litany of calming things that are said to one who might be freaking out. The truly extraordinary thing was that the voice changed as it spoke from stereo to monaural, converging inside her skull, as though she had switched from speakers to headphones.

"It's okay, Sol, I'm all right," she assured him, and then realized that she had not spoken aloud. She tried to and could not.

They drifted for a while in silence, then. And as

they drifted, sparkling darkness everywhere, each became aware of a growing *presence*, for which no words or symbols existed, which their minds could not grasp but only see/feel/taste. Barbara concentrated as hard as she could on the complex abstract which was Solomon Orechal's identity in her mind; received no familiar echo.

Of course, she thought, *of course he sees himself differently than I do*. She waited patiently for her mind to construct a suitable analogy for the identity-waves she was beginning to receive, and wondered what *he* was seeing. *Soon I'll know.*

The darkness coalesced, lightened perceptibly. An image began to take form, seen simultaneously from all angles.

It was a smooth iridium sphere.

It gleamed before her in the swirling dark, self-contained and apparently impenetrable. Her heart began to beat faster, a bass drum miles below her.

As she watched, spellbound, she saw the polished surface of the sphere begin to discolor, to tarnish. Portions of its surface began to bubble and flake away, as though the metallic sphere were immersed in a clear acid that was slowly oxidizing it away. A high, sharp whining became audible, a sound of reluctant disintegration.

The image disturbed and frightened Barbara. She sensed an uncontrollable power latent in the sphere, ready to burst it asunder when it was sufficiently weakened. Girlfriends had tried to tell her of their experiences with Truth, but the closest Barbara had heard to this was a woman who said she initially perceived her partner as a man in full medieval armor, visor down. Unsettled, Barbara found that she was employing a pressure she could not define, in a manner she could not describe, against the sphere she could not understand.

Whatever it is, she screamed silently, *let it end now. It's been too long already.*

Time stood still, and she slipped into a new plane of understanding, intuition refined into knowledge. She perceived all at once that the walls of the sphere drew strength in some way from the marijuana Solomon had smoked—and that he had known they would.

He lied, came the thought.

And at that, the sphere crumbled like a sugar easter egg in a glass of boiling water.

Parts of that explosion of data she forgot as soon as she perceived them. Parts of it she would carry with her to the end of her days. Some things simply could not be forced into words, some translated as paragraphs, some as single words or impressions coded only to subvocalized grunts or wordless cries. Alone in the darkness that crackled and roared she recoiled, struggling to reduce the enormous input to something comprehensible, pursued by howling fragmentary echoes of forgotten thoughts and memories.

. . . thinks he's so smart, I'll break his . . . nobody knows but me . . . so alone like this, I . . . don't look . . . things on so I could squint in the mirror and see what a lady looked like in her . . . don't look in . . . I didn't mean to . . . won't let me, just bec . . . it wasn't cheating exactly, it was . . . don't l . . . so pretty, I wonder what her . . . don't . . . how could she do this to me after all we . . . holy shit, it squirted all over my . . . If only I . . . don't look insi . . . what's he doing to Mother? . . . don't l . . . I . . .

Shaken to her roots, she reeled but held on, too terrified to let go. There was something beneath, something hidden, something that made alarms go off all over her subconscious. And as well as something hidden, there was something missing, and she knew intuitively that they were connected. *What's missing?* she screamed toward the place where she had once supposed God to *be. What is wrong?* The onslaught continued, keeping her off-balance.

Gawd, you give a pain in the ass Janice, you real . . .

think I got away with it this ti . . . got to get a B this term of Old Karkhead'll . . . don't loo . . . God the Father Almighty Who . . . she suspects . . . others kids get a bike so why can't . . . red like blood . . . be good, God, I'll be . . . don't look insi . . . n't you understand I've got to be the master in my own . . . why you . . . don't look . . . seen a . . . sunset . . . like . . . that before . . . hairy black spider that . . . so alone and they . . . don't look inside . . . DON'T LOOK INSIDE! . . .

Inside! With a sinking feeling of terror and despair Barbara yanked her attention from the chaotic distracting turmoil that the sphere had held, and turned it inward. She found only the confusion of her own thoughts.

She was alone inside her skull.

Where was Solomon? Why was he not probing *her* consciousness, as deep within her identity as she was in his?

Frantic now, she reached back out to the welter of tangled thoughts and forgotten memories emanating from her lover, and . . . *swept* at it, in a manner impossible to describe. The roar of swarming images died as though she had struck a suppressor switch, and she saw several things very clearly.

She saw that Solomon had palmed most of his share of the drug.

She saw his consciousness, trembling, crouched, incoherent with terror.

She saw at last that which he had sought most to hide: that the feeling he professed to have for her was nonexistent, a cover for his real motivations.

She saw his true reason for clinging to her: a paralyzing fear of being, in history's most crowded era, intolerably alone.

She saw that her man had never confronted her identity as an individual, never allowed himself to perceive her as a person, as anything but a palliative for hideous loneliness. Nor anyone else in his life.

She saw that he was afraid to confront her identity, to accept the guilt he knew he bore for using another human being as a tool, a teddy-bear, a living fetish with which to ward off demons of solitude.

She saw the indifference with which he regarded her own hopes and needs and fears, saw the relentless guilt which made him despise himself for it.

She saw the desperation in which he had sought to hide the truth from them both by reducing his dosage of yage and distorting both their synaptic responses with pot.

Comprehension and compassion washed over her as a single wave, a wave of pity and love for this tormented man to whom she had given her heart, and she cried out in her mind: *it's all right, Sol, it's ALL RIGHT! Don't be afraid, please. I love you.*

Undrugged, he heard her not.

She saw swimming to the surface of his mind a surreal cartoon figure of herself, choked with revulsion, recoiling from the selfishness of his love, face contorted with bitter rejection. *No!* she screamed silently, but she knew he could not hear, knew she could not make him hear, and knew with astonished horror that he was snapping, could no longer bear the crushing pain of the guilt he could not forget; and she realized with a nauseating certainty what he was going to do.

The throbbing undercurrent of fragmented voices swelled to a shuddering roar in her skull, and now each of those voices was only a throaty growl.

She screamed once, and then many times.

The hissing of the torch reverberated in the bare corridor with an acoustic sibilance that was unpleasant if you listened to it. Jerry and Vito had learned not to listen to it.

"Ain't had this thing out of the shop in so long, I feel like I oughta take it for a walk," Jerry said, adjusting the oxy mix.

"Yeah," grunted Vito from behind his opaque mask.

"Naw, we sure don't have to do this very often."

Vito grunted again.

"Wonder what made him do it. You know? Whole place like that to hisself, nobody to tell him when to go to work, when to go to sleep. Just lie around all day and think about fems, that's what I'd do."

"So get busted," Vito grumbled.

"Hey, bro. What's with you? You got a bellyache or something?"

"Gimme the willies, that bird."

"Him? He ain't givin' nobody nothin'."

Vito grunted a third time, and Jerry shook his head, returning to his cutting. *Welfare check's due tonight*, he thought suddenly, and smiled behind the polarized mask that shielded his eyes from the arc of the torch.

Noises came from the distance, approaching. Hastily, Vito stubbed out a Gold and tucked the roach in his shirt pocket.

The warden came into view around the corner, followed by two long-haired guards. He swept past Vito and Jerry without a word, ignoring the torch, and peered into the window of the cell door.

"Mmmmmmm," he said. "Yes."

The two guards shifted their weight restlessly.

"All right," said the balding official. "All right. Obviously it's a suicide."

"Obviously," murmured one of the guards, a blond, mustached youth. The warden glared at him irritably.

"Why wasn't I notified at once?"

"You were, sir," the guard said evenly. "Union regs say you only have to check 'em twice on night shift unless otherwise ordered. That's how I found him an hour ago. It was already too late to help him."

"Oh, very well, very well," the warden grumbled. "Carry on, you two." He went away, trailing the two guards. The blond one was smiling faintly.

Jerry and Vito looked at each other, shrugged. Jerry

realigned the still-snarling torch against the door, and Vito relit his joint.

"Sure is a good thing this old torch leaks so bad, or he'd have smelled that and taken your ass," Jerry grinned. Vito passed him the joint; he slid it behind his mask and toked quickly, before the smoke could accumulate and lace his eyes. After a time he left off tracing a nearly complete, foot-and-a-half circle in the plasteel door, and paused. Giggling, he began to inscribe eyes and a broadly smiling mouth within the circle. Vito watched and smoked silently.

Again echoes sounded distantly. "Jesus," said Jerry. Vito glared at him and swallowed the joint. Hastily, Jerry completed the circle and began hammering at the disc he had cut, frantic to unseat it before his artwork was seen.

He was barely in time; even as the plug fell into the cell with a crash, two fat men came into view at the end of the corridor. One wore black and one wore gray. Both wore the same expression.

Jerry and Vito scrambled to their feet and backed away from the door, striving to look straight. The fat men came near simultaneously, entirely ignoring the two welfare workers.

The one in gray reached gingerly through the new hole in the cell door, pulled toward himself with a gloved hand. They both entered, walked a few paces inside, stopped.

"Not much either of us can do here, is there, Doctor?"

"It seems not, Father."

"Well, then . . ."

"Yes."

They emerged, began to walk away.

"Hey," Jerry yelped.

The physician turned. "Yes?"

"Wh . . . what do we do with . . . ?"

The fat gray man paused, thought for a moment. "Unlock the infirmary and put him in there somewhere. I'll have a vehicle sent." He and the priest left, talking about chess.

Jerry looked at Vito, who gave him a very black look. He knelt and extinguished the torch, and silence fell in the corridor.

They went inside.

"Jeez," Vito breathed softly. It sounded like a prayer.

The two-inch-thick plug was lying just inside the doorway, its imbecile smile upside down. Beyond it lay McGinny, on his back, a feral and bloody grin on his face. His wrists had been chewed open.

"Jeez," Vito said again, and began putting on his gloves.

Solomon Orechal sat in his chair and surveyed the room which was to be his home for the next twenty-to-life-depending. With a disgusted sigh he picked his J-45 from the bed, hit a G, tuned, hit an E, tuned, hit an E again. Satisfied, he modulated through D back into G, added a seventh.

"*This time next year,*" he sang, and stopped.

After a while he sang "Pack Up Your Sorrows," and that was all right, but when he had finished he found himself wondering who he could give all *his* sorrows to, so he went right into Lightning's "Prison Blues," and managed to get off on that.

But before long, inevitably, he was playing the song he used to close every set, the one he hadn't wanted to play here, now. He was halfway into James Taylor's "Don't Let Me Be Lonely Tonight," when he saw the face at the cell window, blond mustache under a blue uniform hat. He leaped from the chair, tossing his pride-and-joy heedlessly toward the bed, and sprang to the window.

"HEY OUT THERE, can you hear me?" he shouted.

"Hey, man, be measured," came a soft voice, electronically muffled. "I can hear you heavy."

"Wow, listen," Solomon babbled, "you work here, man? Or what? Hey listen, *you want to hear a song?* You got a minute?"

"Sure, bro, sure. Take it easy."

Solomon ran back to the bed, picked up his axe and threw the strap over his head. He began frantically patting his pockets for a flat-pick, discovered he held one in his hand.

"What are you in for?" the blond guard asked quietly.

"Huh? Me? Oh, uh . . . rape," Solomon said, gripping the pick. ". . . and murder," he added, and looked down, hitting a very intricate chord.

The blond jailer's eyes lit up, and he tugged at his mustache.

TIDBIT: interleaf

"Nobody Likes to Be Lonely" is the earliest story in this book, and I am willing to concede, now that you've read it and it's too late, that it is . . . how shall I put this . . . one of the least wonderful. I decided to include it here anyhow: first (and most important) because I am still sneakingly fond of it after all these years; and second because it forms such an interesting contrast with "Satan's Children," the next story.

You will recall that I made a distinction in this book's general introduction between my early and recent stories. I can give you no clearer example than this adjoining pair. They are (dare I say it?) a pair with a great cleavage between them.

"Nobody" was written in 1972, the fifth or sixth story I ever wrote. I wrote it in the editorial department of a businessmen's daily newspaper where I was employed as a liar at the time. That is, I was a real estate editor. My average reader's median income was over fifty thousand per, and my feature stories invariably concerned new advertisers, and it was the first time in my life I had ever made the good coins. I was elegantly miserable. Everyone in the story except Barbara is a creep, and it reeks with cynicism and ends in horror—as did much of my output in those days, come to think.

Six years later I wrote another Truth Dope story.

"Satan's Children" was written less than a year ago, in a happy home fifty feet from the magnificent Bay of Fundy, nestled up against ninety-five acres of woods and heated with the wood I sawed and split myself. I wrote it on the selfsame goddam typewriter. (I bought the old Royal from the newspaper when I quit to go freelance. I keep it with me, to keep me honest.) Everyone in the story bar one is a reasonably nice person, and when I read it aloud at the most recent World Science Fiction Convention, it got *two* ovations: one at the end, and one slightly earlier, when the gun goes off.

And I'm not sure whether it ends on a note of bright optimism . . . or of horror beside which "Nobody" pales in comparison. Or both.

It is never safe to assume that the opinions of a protagonist are those of the author, even if the protagonist is made to seem a reasonably nice person. (This classic error runs through virtually all the Heinlein criticism I've ever seen.) It cannot even be safely assumed that the author *has* any firm opinion on the subject, simply because his characters do.

As I type this interleaf I find that I approve of my protagonist's final decision. But I must admit I am uncomfortably aware that the story could be justifiably retitled, "Events Leading Up To The End Of The World." I am highly suspicious of chemical panaceas, and I really do believe in my heart of hearts that dosing someone without his or her knowledge or consent is ultimate mortal sin (whatever that means); and furthermore, I confess to a perverted affection for the World As We Know It.

But I'd be willing to consider trading it for a better model.

. . . if I could just define "better" in my mind.

Herewith this latest attempt, which begins with murder and ends in bright optimism and horror . . .

6
SATAN'S CHILDREN

A beginning is the end of something, always.

Zaccur Bishop saw the murder clearly, watched it happen—although he was not to realize it for over an hour.

He might not have noticed it at all, had it happened anywhere but at the Scorpio. The victim himself did not realize that he had been murdered for nearly ten minutes, and when he did he made no outcry. It would have been pointless: there was no way to demonstrate that he was dead, let alone that he had been killed, nor anything whatever to be done about it. If the police had been informed—and somehow convinced—of all the facts, they would have done their level best to forget them. The killer was perhaps as far from the compulsive-confessor type as it is possible to be: indeed, that was precisely his motive. It is difficult to imagine another crime at once so public and so clandestine. In any other club in the world it would have been perfect. But since it happened at the Scorpio, it brought the world down like a house of cards.

The Scorpio was one of those clubs that Gods sends every once in a while to sustain the faithful. Benched from the folkie-circuit for reasons he refused to discuss, a musician named Ed Finnegan somehow convinced the owners of a Chinese restaurant near Dalhousie University to let him have their basement and an unreasonable sum of money. (Finnegan used to claim that when he vacationed in Ireland, the Blarney Stone tried to kiss him.) He found that the basement comprised two large windowless rooms. The one just inside the front door he made into a rather conventional bar—save that it was not conventionally overdecorated. The second room, a much larger one which had once held the oil furnace (the building predated solar heat), he painted jet black and ceilinged with acoustic tile. He went then to the University, and to other universities in Halifax, prowling halls and coffeehouses, bars, and dormitories, listening to every musician he heard. To a selected few he introduced himself, and explained that he was opening a club called Scorpio. It would include, he said, a large music room with a proper stage and spotlight. Within this music room, normal human speech would be forbidden to all save the performers. Anyone wishing food or drink could raise their hand and, when the waitress responded, point to their order on the menu silk-screened into the tablecloth. The door to this room, Finnegan added, would be unlocked only between songs. The PA system was his own: six Shure mikes with boomstands, two Teac mixers, a pair of 600-watt Toyota amps, two speaker columns, four wall speakers, and a dependable stage monitor. Wednesday and Thursday were Open Mike Nights, with a thirty-minute-per-act limit, and all other nights were paying gigs. Finnegan apologized for the meagerness of the pay: little more than the traditional all-you-can-drink and hat privileges. The house piano, he added, was in tune.

Within a month the Scorpio was legend, and the Chinese restaurant upstairs had to close at sunset—for lack of parking. There have always been more good serious musicians than there were places for them to play; not a vein for the tapping but an artery. Any serious musician will sell his or her soul for an intelligent, sensitive, *listening* audience. No other kind would put up with Finnegan's house rules, and any other kind was ejected—at least as far as the bar, which featured a free juke box, Irish coffee, and Löwenbräu draft.

It was only because the house rules were so rigidly enforced that Zack happened to notice even that most inconspicuous of murders.

He was about to do the last song in his midevening solo set; Jill sat at a stageside table nursing a plain orange juice and helping him with her wide brown eyes. The set had gone well so far, his guitar playing less sloppy than usual, his voice doing what he wanted it to, his audience responding well. But they were getting restive: time to bottle it up and bring Jill back onstage. While his subconscious searched its files for the right song, he kept the patter flowing.

"No, really, it's true, genties and ladlemen of the audio radiance, I nearly had a contract with Chess Records once. Fella named King came to see me from Chess, but I could see he just wanted old Zack Bishop for his pawn. He was a screaming queen, and he spent a whole knight tryin' to rook me, but finally I says, 'Come back when you can show me a check, mate.'" The crowd groaned dutifully, and Jill held her nose. Lifting her chin to do so exposed the delicate beauty of her throat, the soft grace of the place where it joined her shoulders, and his closing song was chosen.

"No, but frivolously, folks," he said soberly, "it's nearly time to bring Jill on back up here and have her sing a few—but I've got one last spasm in me first. I guess you could say that this song was the proximate cause of Jill and me getting together in the first place. See, I met this lady and all of a sudden it seemed like there was a whole lot of things we wanted to say to each other, and the only ones I could get out of my mouth had to do with, like, meaningful relationships, and emotional commitments, and how our personalities complemented each other and like that." He began to pick a simple C-Em-Am-G cycle in medium slow tempo, the ancient Gibson ringing richly, and Jill

..iled. "But I knew that the main thing I wanted to say had nothing to do with that stuff. I knew I wasn't being totally honest. And so I had to write this song." And he sang:

Come to my bedside and let there be sharing*
Uncounterfeitable sign of your caring
Take off the clothes of your body and mind
Bring me your nakedness . . . help me
     in mine . . .

Help me believe that I'm worthy of trust
Bring me a love that includes honest lust
Warmth is for fire; fire is for burning . . .
Love is for bringing an ending . . .
     to yearning

For I love you in a hundred ways
And not for this alone
But your lovin' is the sweetest lovin'
I have ever known

He was singing directly to Jill, he always sang this song directly to Jill, and although in any other bar or coffeehouse in the world an open fistfight would not have distracted his attention from her, his eye was caught now by a tall, massively bearded man in black leather who was insensitive enough to pick this moment to change seats. The man picked a stageside table at which one other man was already seated, and in the split second glance that Zack gave him, the bearded man met his eyes with a bold, almost challenging manner.

Back to Jill.

Come to my bedside and let there be giving
Licking and laughing and loving and living

*Music and lyrics in Appendix A, p. 311

Sing me a song that has never been sung
Dance at the end of my fingers and tongue

Take me inside you and bring up your knees.
Wrap me up tight in your thighs and then squeeze
Or if you feel like it you get on top
Love me however you please, but please . . .
    don't stop

For I love you in a hundred ways
And not for this alone
But your lovin' is the sweetest lovin'
I have ever known

The obnoxious man was now trying to talk to the man he had joined, a rather elderly gentleman with shaggy white hair and ferocious mustaches. It was apparent that they were acquainted. Zack could see the old man try to shush his new tablemate, and he could see that the bearded man was unwilling to be shushed. Others in the audience were also having their attention distracted, and resenting it. Mentally gritting his teeth, Zack forced his eyes away and threw himself into the bridge of the song.

I know just what you're thinking of
There's more to love than making love
There's much more to the flower than the bloom.
But every time we meet in bed
I find myself inside your head
Even as I'm entering your womb

The Shadow appeared as if by magic, and the Shadow was large and wide and dark black and he plainly had sand. None too gently he kicked the bearded man's chair and, when the latter turned, held a finger to his lips. They glared at each other for a few seconds, and then the bearded man turned around again. He gave up trying to talk to the white-haired man, but Zack had the funny idea that his look of disappointment was counterfeit—he seemed underneath it to be somehow *satisfied* at being silenced. Taking the old man's left hand in his own, he produced a felt-tip pen and began writing on the other's palm. Quite angry now, Zack yanked his attention back to his song, wishing fiercely that he and Jill were alone.

> So come to my bedside and let there be loving
> Twisting and moaning and thrusting and shoving
> I will be gentle—you know that I can
> For you I will be quite a singular man . . .

> Here's my identity, stamped on my genes
> Take this my offering, know what it means
> Let us become what we started to be
> On that long ago night when you first came with me

> Oh lady, I love you in a hundred ways
> And not for this alone
> But your lovin' is the sweetest lovin'
> I have ever known

The applause was louder than usual, sympathy for a delicate song shamefully treated. Zack smiled half-ruefully at Jill, took a deep draught from the Löwenbräu on the empty chair beside him, and turned to deliver a stinging rebuke to the bearded man. But he was gone, must have left the instant the song ended— Shadow was just closing the door behind him. The old man with the absurd mustaches sat alone, staring at the writing on his palm with a look of total puzzlement. Neither of them knew that he was dead. The old man too rose and left the room as the applause trailed away.

To hell with him, Zack decided. He put the beer down at his feet and waved Jill up onto the stage.

"Thank you folks, now we'll bring Jill back up here so she and I can do a medley of our hit . . ."

The set went on.

The reason so many musicians seem to go a little nutty when they achieve success, demanding absurd luxuries and royal treatment, is that prior to that time they have been customarily treated like pigs. In no other branch of the arts is the artist permitted so little dignity by his merchandisers and his audience, given so little respect or courtesy. Ed Finnegan was a musician himself, and he understood. He knew, for instance, that a soundproof dressing room is a pearl without price to a musician, and so he figured out a cheap way to provide one. He simply erected a single soundproof wall, parallel to the music room's east wall and about five feet from it. The resulting corridor was wide enough to allow two men with guitars to pass each other safely, long enough to pace nervously, and silent enough to tune up or rehearse in.

And it was peaceful enough to be an ideal place to linger after the last set, to recover from the enormous expenditure of energy, to enjoy the first *tasted* drink of the night, to hide from those dozens of eyes half-seen through spotlight glare, to take off the sweat-soaked image and lounge around in one's psychic underwear. The north door led to the parking lot and was always locked from the outside; the south door opened onto stage right, and had a large sign on its other side that said clearly, "If the performers wish to chat, sign autographs, accept drinks or tokes or negotiate for your daughter's wedding gig, they will have left this door open and you won't be reading this. PLEASE DO NOT ENTER. DON'T KNOCK IF YOU CAN HELP IT. RESPECT US AND WE'LL MAKE BETTER MUSIC. Thank you—Finnegan."

It was sanctuary.

Zack customarily came offstage utterly exhausted, while Jill always finished a gig boiling over with nervous energy. Happily, this could be counterbalanced by their differing metabolic reactions to marijuana: it always gave Zack energy and mellowed Jill. The after-gig toke was becoming a ritual with them, one they looked forward to unconsciously. Tonight's toke was a little unusual. They were smoking a literal cigar of grass, GMI's newest marketing innovation, and assessing the validity of the product's advertising slogan: "It doesn't get you any higher—but it's more fun!"

Zack lay on his back on the rug, watching excess smoke drift lazily up from his mouth toward the high ceiling. An internal timer went off and he exhaled, considered his head. "Let me see that pack," he said, raising up on one elbow. Jill, just finishing her own toke, nodded and passed over both cigar and pack.

Zack turned the pack over, scanned it and nodded. "Brilliant," he said. He was beginning to come out of his postperformance torpor. He toked, and croaked "Fucking brilliant" again.

Jill managed to look a question while suppressing a cough.

He exhaled. "Look," he said. "'Guaranteed 100% pure marijuana.' See what that means?"

"It means I'm not crazy, I really *am* stoned."

"No, no, the whole cigar business. Remember the weather we had last spring? Half the GMI dope fields got pasted with like thirty-two straight days of rain, which is terrific for growing rope and rotten for growing smoke. Stalks like bamboo, leaves like tiny and worth squat, dope so pisspoor you'd have to smoke a cigar-full of it to get off. So what did they do?" He grinned wolfishly. "*They made cigars.* They bluffed it out, just made like they planned it and made cigars. They're pure grass, all right—but you'd have to be an idiot to smoke a whole cigar of *good* grass. And by Christ I'll bet they pick up a big share of the market. These things *are* more fun."

"What do you think that is?" Jill asked. "*Why* is it more fun? Is it just the exaggerated oral trip?"

"Partly that," he admitted. "Oh hell, back when I smoked tobacco I knew that cigars were stronger, cooler, and tastier—I just couldn't afford 'em. But these aren't much more expensive than joints. Breaks down to about a dime a hit. Why, don't you like 'em?"

She took another long toke, her expression going blank while she considered. Suddenly her eyes focused, on him. "Does it turn you on to watch me smoke it?" she asked suddenly.

He blushed to his hairline and stammered.

"Honesty, remember? Like you said when you sang our song tonight. Trust me enough to be honest."

"Well," he equivocated, "I hadn't thought about . . ." He trailed off, and they both said "bullshit" simultaneously and broke up. "Yeah, it turns me on," he admitted.

She regarded the cigar carefully, took a most sensuous toke. "Then I shall chain smoke 'em all the way home," she said. "Here." She handed him the stogie, then began changing out of her stage clothes, making a small production out of it for him.

Eight months we've been living together, Zack thought, *and she hasn't lost that mischievous enthusiasm for making me horny. What a lady!* He put the cigar in his teeth, waggled it and rolled his eyes. "Why wait 'til we get home?" he leered.

"I predict another Groucho Marx revival if those things catch on." Her bra landed on top of the blouse.

"I like a gal with a strong will," he quoted, "or at least a weak won't." He rose and headed for her. She did not shrink away—but neither did she come alive in his arms.

"Not here, Zack."

"Why not? It was fun in that elevator, wasn't it?"

"That was different. Someone could come in."

"Come on, the place is closed, Finnegan and the Shadow are mopping up beer and counting the take, nobody's gonna *fuck* a *duck*."

Startled, she pulled away and followed his gaze. A shining figure stood in the open doorway.

She was by now wearing only ankle-length skirt and panties, and Zack had the skirt halfway down her hips, but she, and he, stood quite still, staring at the apparition. It was several moments after they began wishing for the power of motion that they recalled that they possessed it; moments more before they used it.

"It was true," the old man said.

He seemed to shine. He shimmered, he crackled with an energy only barely invisible, only just intangible. His skin and clothes gave the impression of being on the verge of bursting spontaneously into flame. He shone as the Christ must have shone, as the Buddha must have shone, and a Kirlian photograph of him at that moment would have been a nova-blur.

Zack had a sudden, inexplicable and quite vivid recollection of the afternoon of his mother's funeral, five years past. He remembered suddenly the way friends and relatives had regarded him as strangers, a little awed, as though he possessed some terrible new power. He remembered feeling at the time that they were correct—that by virtue of his grief and loss he was somehow charged with a strange kind of energy. Intuitively he had *known* that on this day of all days he could simply scream at the most determined and desperate mugger and frighten him away, on this day he could violate traffic laws with impunity, on this day he could stare down any man or woman alive. Coming in close personal contact with death had made him, for a time, a kind of temporary shaman.

And the old man was quite dead, and knew it.

"Your song, I mean. It was true. I was half afraid I'd find you two bickering, that all that affection was just a part of the act. Oh thank *God*."

Zack had never seen anyone quite so utterly relieved. The old man was of medium height and appeared to be in robust health. Even his huge ungainly mustaches could not completely hide the lines of over half a century of laughter and smiles. His complexion was ruddy, his features weatherbeaten, and his eyes were infinitely kindly. His clothes were of a style which had not even been revived in years: bell bottom jeans, multicolored paisley shirt with purple predominating, a double strand of beads and an Acadian scarf-cap sloppily tied. He wore no jewelry other than the beads and no makeup.

A kind of Hippie Gepetto, Zack decided. *So why am I paralyzed?*

"Come in," Jill said, and Zack glanced sharply at her, then quickly back. The old man stepped into the room, leaving the door ajar. He stared from Zack to Jill and back again, from one pair of eyes to the other, and his own kindly eyes seemed to peel away onion-layers of self until he gazed at their naked hearts. Zack suddenly wanted to cry, and that made him angry enough to throw off his trance.

"It is the custom of the profession," he said coldly, "to knock and shout, 'Are you decent?' Or didn't you see the sign there on the door?"

"Both of you are decent," the old man said positively. Then he seemed to snap out of a trance of his own: his eyes widened and he saw Jill's half-nakedness for the first time. "*Oh,*" he said explosively, and then his smile returned. "Now I'm supposed to apologize," he twinkled, "but it wouldn't be true. Oh, I'm sorry if I've upset you—but that's the last look I'll ever get, and you're lovely." He stared at Jill's bare breasts for a long moment, watched their nipples harden, and Zack marveled at his own inability to muster outrage. Jill just stood there . . .

The old man pulled his eyes away. "Thank you both. Please sit down, now, I have to say some preposterous things and I haven't much time. Please hear me out before you ask questions, and please—please!—believe me."

Jill put on the new blouse and jeans, while Zack

seated himself from long habit on the camel-saddle
edge of his guitar case. He was startled to discover the
cigar still burning in his hand, stunned to see only a
quarter-inch of ash on the end. He started to offer it
to the old man; changed his mind; started to offer it to
Jill; changed his mind; dropped the thing on the carpet
and stepped on it.

"My name is Wesley George," the old man began.

"Right," Zack said automatically.

The old man sighed deeply. "I haven't much time,"
he repeated.

"What" *the hell would Wesley George be doing in
Halifax?* Zack started to say, but Jill cut him off
sharply with "He's Wesley George and he doesn't have
much time" and before the intensity in her voice he
subsided.

"Thank you," George said to Jill. "You *perceive* very
well. I wonder how much you know already."

"Almost nothing," Jill said flatly, "but I know what
I know."

He nodded. "Obviously you've both heard of me;
Christ knows I'm notorious enough. But how much of
it stuck? Given my name, how much do you know of
me?"

"You're the last great dope wizard," Zack said, "and
you were one of the first. You used to work for one of
the 'ethical' drug outfits and you split. You synthesized DMT, and didn't get credit for it. You developed
Mellow Yellow. You made STP safe and dependable.
You develop new psychedelics and sell 'em cheap, sometimes you give 'em away, and some say you're stone
nuts and some say you're the Holy Goof himself. You
followed in the footsteps of Owsley Stanley, and you've
never been successfully busted, and you're supposed to
be richer than hell. A dealer friend of mine says you
make molecules talk."

"You helped buy the first federal decrim bill on
grass," Jill said, "and blocked the cocaine bill—both
from behind the scenes. You founded the Continent

Continent movement and gave away five million TM pills in a single day in New York."

"Some people say you don't exist," Zack added.

"As of now, they're right," George said. "I've been murdered."

Jill gasped; Zack just stared.

"In fact, you may have noticed it done," Wesley said to Zack. "You remember Sziller, the bearded man who spoiled your last solo? Did you see him write this?"

George held his left hand up, palm out. A black felt-tip pen had written a telephone number there, precisely along his life-line.

"Yeah," Zack agreed. "So what?"

"I dialed it a half hour ago. David Steinberg answered. He said that once he had a skull injury, and the hospital was so cheap they put a *paper* plate in his head. He said the only side effect was that every sunny day he *had* to go on a picnic. I hung up the phone and I knew I was dead."

"Dial-a-Joke," Jill said wonderingly.

"I don't get it," said Zack.

"I was supposed to meet Sziller here tonight—in the bar, after your set. I couldn't understand why he came into the music room and tried to talk to me there. He knew better. He *wanted* to be shushed, so he'd have to write his urgent message on my hand. And the urgent message was literally a joke. So what he really wanted was to write on my hand with a felt-tip pen."

"Jesus," Zack breathed, and Jill's face went featureless.

"In the next ten or fifteen minutes," George said conversationally, "I will have a fatal heart attack. It's an old CIA trick. A really first-rate autopsy might pick up some traces of a phosphoric acid ester—but I imagine Sziller and his people will be able to prevent that easily enough. They've got the building surrounded; I can't get as far as my car. You two are my last hope."

Zack's brain throbbed, and his eyelids felt packed

with sand. George's utter detachment was scary. It said that Wesley George was possessed by something that made his own death unimportant—and it might be catching. His words implied that it was, and that he proposed to infect Zack and Jill. Zack had seen *North By Northwest*, and had no intention of letting other peoples' realities hang him out on Mount Rushmore if there were any even dishonorable way to dodge.

But he could *perceive* pretty well himself, and he knew that whatever the old man had was a burden, a burden that would crush him even in death unless he could discharge it. Everything that was good in Zack yearned to answer the call in those kindly eyes; and the internal conflict—almost entirely subconscious—nearly tore him apart.

There was an alternative. It would be easy to simply disbelieve the old man's every word. Was it plausible that this glowing, healthy man could spontaneously die, killed by a bad joke? Zack told himself that Hitler and Rasputin had used just such charisma to sell the most palpable idiocies, that this shining old man with the presence of a Buddha was only a compelling madman with paranoid delusions. Zack had never seen a picture of Wesley George. He remembered the fake Abby Hoffman who had snarled up the feds for so long. He pulled skepticism around himself like a scaly cloak, and he looked at those eyes again, and louder and more insistent even than Jill's voice had been, they said that the old man was Wesley George and that he didn't have much time.

Zack swallowed something foul. "Tell us," he said, and was proud that his voice came out firm.

"You understand that I may get you both dipped in soft shit, maybe killed?"

Zack and Jill said, "Yes," together, and glanced at each other. This was a big step for both of them: there is all the difference in the world between agreeing to live together and agreeing to die together. Zack knew that whatever came afterward, they were married as of now, and he desperately wanted to think that through, but there was no time, no time. *What's more important than death and marriage?* he thought, and saw the same question on Jill's face, and then they turned back as one to Wesley George.

"Answer me a question first," the old man said. Both nodded. "Does the end justify the means?"

Zack thought hard and answered honestly. Much, he was sure, depended on this.

"I don't know," he said.

"Depends on the end," Jill said. "And the means."

George nodded, content. "People with a knee jerk answer *either* way make me nervous," he said. "All right, children, into your hands I place the fate of modern civilization. I bring you Truth, and I think that the truth shall make you flee."

He glanced at his watch, displayed no visible reaction. But he took a pack of tobacco cigarettes from his shirt, lit one, and plainly gave his full attention to savoring the first toke. Then he spoke, and for the first time Zack noticed that the old man's voice was a pipe organ with a double bass register, a great resonant baritone that Disraeli or Geronimo might have been proud to own.

"I am a chemist. I have devoted my life to studying chemical aspects of consciousness and perception. My primary motivation has been the advancement of knowledge; my secondary motive has been to get people high—as many people, as many ways as possible. I think the biggest single problem in the world, for almost the last two decades, has been morale. Despairing people solve no problems. So I have pursued better living through chemistry, and I've made my share of mistakes, but in the main I think the world has profited from my existence as much as I have from its. And now I find that I am become Prometheus, and that my friends want me dead just as badly as my enemies.

"I have synthesized truth.

SATAN'S CHILDREN 203

"I have synthesized truth in my laboratory. I have distilled it into chemical substance. I have measured it out in micrograms, prepared a dozen vectors for its use. It is not that hard to make. And I believe that if its seeds are once sown on this planet, the changes it will make will be the biggest in human history.

"Everything in the world that is founded on lies may die."

Zack groped for words, came up empty. He became aware that Jill's hand was clutching his tightly.

"'What is truth?' asked jesting Pilate, and would not stay for answer. Neither will I, I'm afraid—but I ought to at least clarify the question. I cannot claim to have objective truth. I have no assurance that there is such a thing. But I *have* subjective truth, and I *know* that exists. I knew a preacher once who got remarkable results by looking people square in the eye and saying, 'You do *too* know what I mean.'"

A spasm crossed the old man's face and his glowing aura flickered. Zack and Jill moved toward him as one, and he waved them away impatiently.

"Even those of us who pay only lip service to the truth know what it is, deep down in our hearts. And we all believe in it, and know it when we see it. Even the best rationalization can fool only the surface mind that manufactures it; there is something beneath, call it the heart or the conscience, that knows better. It tenses up like a stiff neck muscle when you lie, in proportion to the size of the lie, and if it stiffens enough it can kill you for revenge. Ask Richard Corey. Most people seem to me, in my cynical moments, to keep things stabilized at about the discomfort of a dislocated shoulder or a tooth about to abscess. They trade honesty off in small chunks for pleasure, and wonder that their lives hold so little *joy*. Joy is incompatible with tensed shoulders and a stiff neck. You become uneasy with people in direct proportion to how many lies you have to keep track of in their presence.

"I have stumbled across a psychic muscle relaxant."

"Truth serum's been around a long time," Zack said.

"This is no more pentothal than acid is grass!"
George thundered, an Old Testament prophet enraged. He caught himself at once—in a single frantic instant he seemed to extrude his anger, stare at it critically, tie it off, and amputate it, in deliberate steps. "Sorry—rushed. Look: pentothal will—sometimes—get you a truthful answer to a direct question. *My* drug imbues you with a strong desire to get straight with all the people you've been lying to, regardless of consequences. Side effects include the usual accompaniments of confession—cathartic relief, euphoria to the point of exaltation and a tendency to babble—and a new one: visual color effects extremely reminiscent of organic mescaline."

He winced again, clamped his jaw for a moment, then continued.

"That alone might have been enough to stand the world on its ear—but the gods are jollier than that. The stuff is water-soluble—damn near anything-soluble—and skin-permeable and as concentrated as hell. Worse than acid for dosage, and it can be taken into the body just about every way there is. For pentothal you have to actually shoot up the subject, and you have to hit the vein. My stuff—Christ, you could let a drop of candle wax harden on your palm, put a pinpoint's worth on the wax, shake hands with a man and dose him six or seven hours' worth. You could put it on a spitball and shoot it through a straw. You could add it to nail polish or inject it into a toothpaste tube or roll it up in a joint or simply spray it from an atomizer. Put enough of it into a joint, in a small room, and even the nonsmokers will get off. The method Sziller has used to assassinate me would work splendidly. There may be some kind of way to guard against it—some antidote or immunization—but I haven't found it yet. You see the implications, of course."

At some point during George's speech Zack had

reached the subconscious decision to believe him implicitly. With doubt had gone the last of his paralysis, and now his mind was racing faster than usual to catch up. "Give me a week and a barrel of hot coffee and I think I could reason out most of the major implications. All I get now is that you can make people be truthful against their will." His expression was dark.

"Zack, I know this sounds like sophistry, but that's a matter of definition. Whoa!" He held up his hands. "I know, son, I know. The Second Commandment of Leary: 'Thous shalt not alter thy brother's consciousness without his consent.' So how about retroactive consent?"

"Say again."

"The aftereffects. I've administered the drug to blind volunteers. They knew only that they were sampling a new psychedelic of unknown effect. In each case I gave a preliminary 'attitude survey' questionnaire with a few buried questions. In fourteen cases I satisfied myself that the subject would probably *not* have taken the drug if he or she had known its effect. In about three-quarters of them I damn well knew it.

"The effects were the same for all but one. All fourteen of them experienced major life upheaval—usually irreversible and quite against their will—while under the effects of the drug. They all became violently angry at me after they came down. Then all fourteen stormed off to try and put their lives back together. Thirteen of them were back within a week, asking me to lay another hit on them."

Zack's eyes widened. "Addictive on a single hit. Jesus."

"No, *no!*" George said exasperatedly. "It's not the drug that's addictive, dammit. *It's the truth that's addictive.* Every one of those people came back for, like, three-four hits, and then they stopped coming by. I checked up on the ones I was in a position to. They had just simply rearranged their lives on solid principles of truth and honesty and begun to live that way all the time. *They didn't need the drug anymore.* Every damn one of them thanked me. One of them fucked me, sweetly and lovingly—at my age.

"I was worried myself that the damned stuff might be addictive—after all, I'm always my own first guinea pig. So I had at least as many subjects who *would* probably have taken the drug knowingly, and *all* of them asked for more and I told them no. Better than three-fourths of them have made similar life adjustments on their own, without any further chemical aid.

"Zack, living in truth *feels good.* And it sticks in your memory. Like, it's a truism with acid heads that you can never *truly* remember what tripping feels like. You *think* you do, but every time you trip it's like waking up all over again, you recognize the head coming on and you dig that your memories of it were shadows. *But this stuff you remember!* You're left with a vivid set of memories of just exactly how good it felt to not have any psychic muscles bunched up for the first time since you were two years old. You remember joy; and you realize that you can recreate it just by not ever lying any more. That's goddam hard, so you look for any help you can get, and if you can't get any you just take your best shot.

"Those people ended up happier, Zack.

"Zack, Jill . . . a long long time ago a doctor named Watt slapped me on the ass and forced me to live. It was very much against my will; I cried like hell and family legend says I tried to bite him. Now my days are ended, and taking it all together I'm very glad he went to the trouble. He had my retroactive consent. It wasn't his fault anyhow: my parents had already forced me to exist, before I had a will for it to be against—and they have my retroactive consent. Many times in my life, good friends and even strangers have kicked my ass where it needed kicking; at least twice women have gently and compassionately kicked

me out—all against my will, and they all have my retroactive consent, god bless 'em. Can it be immoral to dose
folks if you get no complaints?"

"What about the fourteenth person?" Jill asked.
"The one who didn't come back?"

George grimaced. "Touché."

"Beg pardon?"

"Nothing's perfect. The fourteenth man killed me."

"*Oh.*"

The temperature in the room was moderate, but
George was drenched with sweat; his rubby complexion was paling rapidly.

"Look, you two make up your own minds. You can
help them haul me out to the ambulance in a few minutes and then walk away and forget you ever met me,
if that's what you want. But I have to ask you: please,
take over this karma for me. Someone has to, one way
or the other: I seriously doubt that the drug will ever
be found again."

"Is there like a set of instructions for the stuff?"
Zack asked. *Involved,* his head told him. "Notes,
molecule diagrams—" *Somebody's getting infucking-
volved* . . .

"Complete instructions for synthesis, and about ten
liters of the goods, in various forms. That's about
enough to give everything on earth with two legs a
couple of hits apiece. I tell you, it's easy to make. And
it's *fucking* hard to stumble across. If I die, it dies
with me, maybe forever. Blind luck *I* found it, just
blind—"

"Where?" Zack and Jill interrupted simultaneously.

"Wait a minute, you've got to understand. It's in a
very public place—I thought that was a good idea at
the time, but . . . never mind. The point is, from the
moment you pick up the stuff, you must be very very
careful. They don't have to physically touch you—try
not to let anyone come near you if you can help it,
anyone at all—"

"I'll know a fed when I see one," Zack said grimly, "north *or* south."

"No, NO, not feds, not *any* kind of feds! Think that way and you're dead. It wasn't feds that killed me."

"Who then?" Jill asked.

"In my line of work, I customarily do business with a loosely affiliated organization of non-Syndicate drug dealers. It has no name. It is international in scope, and if it ever held a meeting, a substantial fraction of the world's wealth would sit in one room. I offered them this drug for distribution, before I really understood what I had. Sziller is one of the principals of the group."

"Jesus God," Zack breathed. "*Dealers* had *Wesley George* snuffed? That's like the apostles offing Jesus."

"One of them did," George pointed out sadly. "Think it through, son: dope dealers can't afford honesty."

"But—"

"Suppose the feds did get hold of the stuff," Jill suggested.

"Oh."

"Or the Syndicate," George agreed. "Or their own customers, or—"

"What's the drug called?" Jill asked.

"The chemical name wouldn't mean a thing to some of the brightest chemists in the world, and I never planned to market it under that. Up until I knew what it was I called it The New Batch, and since then I've taken to calling it TWT. The Whole Truth." Suddenly urgency overtook him and he was angry again. "Listen, fuck this," he blazed, "I mean fuck all this garbage. OK? I haven't got time to waste on trivia. Will you do it is the important thing; will you take on the karma I've brought you? Will you turn Truth loose on the world for me? Please, you aaaAAAAAAHHH-EE shit." He clutched at his right arm, screamed again in awful pain and fell to the floor.

"We promise, we promise," Jill was screaming, and

Zack was thundering "Where, *where?* Where, dammit?" and Jill had George's head on her lap and Zack had his hands and they clutched like steel and "*Where?*" he shouted again, and George was bucking in agony, breathing in with great whooping gargles and breathing out with sprays of saliva, jaw muscles like bulging biceps on his face, and "Hitch" he managed through his teeth, and Zack tried, "Hitch. Hitchhike, *a locker at the hitching depot*" very fast and then added "Key in your pocket?" and George borrowed energy from his death struggle to nod twice, "Okay, right Wesley, it's covered, man," and George relaxed all over at once and shat his pants. They thought he was dead, then, but blue-gray eyelids rolled heavily up one last time and he saw Jill's face over his, raining tears. "Nice tits . . ." he said. ". . . Thanks . . . children . . . thanks . . . sorry," and in the middle of the last word he did die, and his glowing aura died with him.

The Shadow was standing in the doorway, filling it full, breathing hard. "I heard the sound, man, what—oh holy *shit,* man. What the fuck *happenin'* here?"

Zack's voice was perfect, his delivery impeccable, startled but not involved. "What can I tell ya, Shadow? The old guy comes back to talk blues and like that and his pump quits. Call the croaker, will ya? And pour me a triple."

"Shee-it," the Shadow rumbled. "Nev' a dull night aroun' this fuckin' joint. Hey, *Finnegan!* Finnegan, god damn it." The big black bouncer left to find his boss.

Zack found a numbered key in George's pants, and turned to Jill. Their eyes met and locked. "Yes," Jill said finally, and they both nodded. And then together they pried Zack's right hand from the clutching fingers of the dead dope wizard, and together they made him comfortable on the floor, and then they began packing up their instruments and gear.

* * *

Zack and Jill held a hasty war council in the flimsy balcony of their second-floor apartment. It overlooked a yard so small it would have been hard put not to, as Zack loved to say, and offered a splendid view of the enormous oil refining facility across the street. The view of Halifax Harbor which the architect had planned was forever hidden now behind it, but the cooling breezes still came at night, salt-scented and rich. Even at 2 A.M. the city was noisy, like a dormitory after lights out, but all the houses on this block were dark and still.

"I think we should pack our bags," Zack said, sipping coffee.

"And do what?"

"The dealers must know that Wesley brought a large amount of Truth with him—he intended to turn it over to them for distribution. They don't know *where* it's stashed, and they must be shitting a brick wondering who else does. We're suspect because we're known to have spoken with him, and a hitching depot is a natural stash—so we don't go near the stuff."

"But we've *got* to—"

"We will. Look, tomorrow we're *supposed* to go on tour, right?"

"Screw the tour."

"No, hon, look! This is the smart way. We do just exactly what we would have done if we'd never met Wesley George. We act natural, do the tour as planned—we pack our bags and *go down to the hitching depot* and take off. But some friend of ours—say, John—goes in just ahead of us and scores the bag. Then we show up and ignore him, and by and by the three of us make up a full car for somebody, and after we're out of the terminal and about to board, out of the public eye, John changes his mind and fades and we take over the bag. Zippo bang, off on tour."

"I'll say it again. Screw the tour. We've got more important things to do."

"Like what?"

"?"

"What do *you* wanna do with the stuff? Call the reporters? Stand on Barrington Street and give away samples? Call the heat? Look. We're proposing to unleash truth on the world. I'm willing to take a crack at that, but I'd like to live to see what happens. So I don't want to be connected with it publicly in any way if I can help it. We keep our cover and do our tour—and we sprinkle fairy dust as we go."

"Dose people, you mean?"

"Dose the most visible people we can find, and make damn sure we don't get caught at it. We're supposed to hit nineteen cities in twenty-eight days, in a random pattern that even a computer couldn't figure out. I intend to leave behind us the god-damndest trail of headlines in history."

"Zack, I don't follow your thinking."

"Okay." He paused, took a deep breath, slowed himself visibly. "Okay . . . considering what we've got here, it behooves me to be honest. I have doubts about this. Heavy doubts. The decision we're making is incredibly arrogant. We're talking about destroying the world, as we know it."

"To hell with the world as we know it, Zack, it stinks. A word of truth *has* to be better."

"Okay, in my gut something agrees with you. But I'm still not sure. A world of truth may be better—but the period of turmoil while the old world collapses is sure going to squash a lot of people. Nice people. Good people. Jill, something *else* in my gut suspects that *maybe even good people need lies sometimes.*

"So I want to hedge my bets. I want to experiment first and see what happens. To do that I have to make another arrogant decision: to dose selected individuals, cold-bloodedly and without giving them a chance, let alone a vote. Wesley experimented himself, with a lab and volunteers and procedure and tests, until he proved to his satisfaction that it was okay to turn this stuff loose. Well, I haven't got any of that—but I have to establish to *my* satisfaction that it's cool."

"Do you doubt his results?" Jill asked indignantly.

"To *my* satisfaction. Not Wesley's, or even yours, my darling, or anyone else's. And yes, frankly, I have some doubts about his results."

Jill clouded up. "How can you—"

"Baby, *listen to me.* I believe that every word Wesley George said to us was the absolute unbiased truth as he knew it. But *he himself had taken the drug.* That makes him suspect."

Jill dropped her eyes. "That retroactive consent business bothered me a little too."

Zack nodded. "Yeah. If everybody comes out of prefrontal lobotomy with a smile on his face, what does that prove? If you kidnapped somebody and put a droud in their head, made 'em a wirehead, they'd thank you on their way out—but so what? Things like that are like scooping out somebody's *self* and replacing it with a new one. The new one says thanks—but the old one was *murdered.* I want to make *sure* that Homo Veritas is a good thing—*in the opinion of homo sapiens.*

"So I propose that neither of us take the drug. I propose that we abstain, and take careful precautions not to accidentally contaminate ourselves while we're using it. We'll dose others but not ourselves, and then when the tour is over—or sooner if it feels right—we'll sit down and look over what we've done and how it turned out. Then if we're still agreed, we'll take a couple of hits together and call CBC News. By then there'll be so much evidence they'll have to believe us, and then . . . then the word will be out. Too far out for the dealers to have it squashed or discredited. Or the government."

"And then the world will end."

"And a new one will begin . . . but first we've got to *know.* Am I crazy or does that make sense to you?"

Jill was silent a long time. Her face got the blank

look that meant she was thinking hard. After a few minutes she got up and began pacing the apartment. "It's risky, Zack. Once the headlines start coming, they'll figure out what happened and come after us."

"And the only people who know our schedule are Fat Jack and the Agency. We'll tell 'em there's a skip tracer after us and they'll both keep shut—"

"But—"

"Jill, this ain't the feds after us—it's a bunch of dealers who dasn't let anybody know they exist. They *can't* have the resources they'd need to trace us, even if they did know what city we were in."

"They might. A dealers' union'd have to be international. That's a lot of weight, Zack, a lot of money."

"Darlin'—if all you got is pisspoor dope . . ." He broke off and shrugged.

Jill grinned suddenly. "You make cigars. Let's get packed. More coffee?"

They took little time in packing and preparing their apartment for a long absence. This would be their third tour together; by now it was routine. At last everything that needed doing was done, the lights were out save for the bedside lamp, and they were ready for bed. They undressed quickly and silently, with no flirting byplay, and slid under the covers. They snuggled together spoon fashion for a few silent minutes, and then Zack began rubbing her neck and shoulders with his free hand, kneading with guitarist's fingers and lover's knowledge. They had not yet spoken a word of the change that the events of the evening had brought to their relationship, and both knew it, and the tension in the room was thick enough to smell. Zack thought of a hundred things to say, and each one sounded stupider than the last.

"Zack?"

"Yeah?"

"We're probably going to die, aren't we?"

"Were positively going to die." She stiffened almost

imperceptibly under his fingers. "But I could have told you that yesterday, or last week." She relaxed again. "Difference is, yesterday I couldn't have told you positively that we'd die *together*."

Zack would have sworn they were inextricably entwined, but somehow she rolled round into his arms in one fluid motion, then pulled him on top of her with another. Their embrace was eight-legged and whole-hearted and completely nonsexual, and about a minute of it was all their muscles would tolerate. Then they drew apart just far enough to meet each other's eyes. They shared that, too, for a long minute, and then Zack smiled.

"Have you ever noticed that there is no position or combination of positions in which we do *not* fit together like nesting cups?"

She giggled, and in the middle of the giggle tears leaked from her laughing eyes. "Oh, Zack," she cried, and hugged him again. "I love you so much."

"I know, baby, I know," he murmured in her ear, stroking her hair. "It's not every day that you find something worth dying for—*and* something worth living for. Both at the same time. Christ, I love you."

They both discovered his rigid erection at the same instant, and an instant later they discovered her sopping wetness, and for the first time in their relationship their loins joined without manual aid from either of them. Together they sucked air slowly through their teeth, and then he began to pull his head back to meet her eyes and she stopped him, grabbing his head with her hands and pushing her tongue into his ear. His hips arched reflexively, his hands clutched her shoulders, her legs locked round his, and the oldest dance began again. It was 11 A.M. before they finally slept, and by that time they were in someone else's car, heading, ironically enough, north by northwest.

It's the best way out of Halifax.

* * *

The reader wishing a detailed account of Zack and Jill's activities over the next month can find it at any library with a good newstape and newspaper morgue. The reader is advised to bring a lunch. At any time of year the individual stories that the two folksingers sowed behind them like depth charges would have been hot copy—but God had ordained that Wesley George drop dead in August, smack in the middle of the Silly Season. The news media of the entire North American Confederation went into grateful orgasmic convulsions.

Not all the stories made the news. The events involving the Rev. Schwartz in Montreal, for instance, were entirely suppressed at the time, by the husbands involved, and have only recently come to light. When militant radical leader Mtu Zanje, the notorious "White Mau Mau," was found in Harlem with bullets from sixteen different unregistered guns in him, there was at that time nothing to connect it with the other stories, and it got three inches on page forty-three.

Indeed, the most incredible thing in retrospect is that no one, at the time, connected *any* of the stories. Though each new uproar was dutifully covered in detail, not one journalist, commentator or observer divined any common denominator in them until the month was nearly up. Confronted with the naked truth, the people of North America did not recognize it.

But certainly every one of them saw it or heard about it, in living color stereo and thirty-six point type and four-channel FM, in weekly news magazines and on documentary shows, in gossip columns and radio talk shows, in political cartoons and in comedians' routines. Zack and Jill strongly preferred to examine their results from a distance, and so they tended to be splashy.

In St. John, New Brunswick, they hit an elderly and prominent judge who had more wrinkles than a William Goldman novel, while he was sitting in open court on a controversial treason case. After an astonishing twenty-seven-minute monolog, the aged barrister died in a successful attempt to cover, with the sidearm he had snatched from his bailiff, the defendant's escape. Zack and Jill, sitting in the audience, were considerably startled, but they had to agree that only once had they seen a man die happier: the judge's dead face was as smooth as a baby's.

In Montreal (in addition to the Rev. Schwartz), they managed to catch a Conservative MP on his way into a TV studio and shake his hand. The program's producer turned out to have seen the old movie *Network*—he kept the politician on the air, physically knocking down the programming director when that became necessary. The MP had been—er—liberally dosed; after forty-five minutes of emotional confession he began specifically outlining the secret dreams he had had ever since he first took office, the really *good* programs he had constructed in his imagination but never dared speak aloud, knowing they could never be implemented in the real world of power blocs and interest groups. He went home that night a broken but resigned man, and woke up the next morning to confront a landslide of favorable response, an overwhelming mandate to implement his dreams. To be sure, very very few of the people who had voted for him in the last election ever did so again. But in the *next* election (and every subsequent election involving him) the ninety percent of the electorate who traditionally never vote turned out almost to a person. The producer is now his chief aide.

In Ottawa they tried for the Prime Minister, but they could not get near him or near anything that could get near him. But they did get the aging Peter Gzowski on *90 Minutes Live*. He too chanced to have seen *Network*, and he had much more survival instinct than its protagonist: the first thing he did upon leaving the studio was to make an extensive tape recording and mail several dubs thereof to friends with instructions for their disposal in the event of his sudden death. Accordingly he is still alive and broadcasting today, and there are very few lids left for him to tear off these days.

Outside Toronto Zack and Jill made their most spectacular single raid, at the Universal Light and Truth Convocation. It was a kind of week-long spiritual olympics: over a dozen famous gurus, swamis, reverends, Zen masters, Sufis, priests, priestesses and assorted spiritual teachers had gathered with thousands of their followers on a donated hundred-acre pasture to debate theology and sell each other incense, with full media coverage. Zack and Jill walked through the Showdown of the Shamen and between them missed not a one. One committed suicide. One went mad. Four denounced themselves to their followers and fled. Seven denounced themselves to their followers and stayed. Four wept too hard to speak, the one the others called The Fat Boy (although he was middle-aged) bit off his tongue, and exactly one teacher—the old man who had brought few followers and nothing for sale—exhibited no change whatsoever in his manner or behavior but went home very thoughtfully to Tennessee. It is now known that he could have blown the story then and there, for he was a telepath, but he chose not to. The single suicide bothered Jill deeply; but only because she happened to know of and blackly despise that particular holy man, and was dismayed by the pleasure she felt at his death. But Zack challenged her to name one way in which his demise either diminished the world or personally benefited her, and she came tentatively to accept that her pleasure might be legitimate.

They happened to arrive in Detroit just before the annual meeting of the Board of Directors of General Motors. Madame President absent-mindedly pocketed the cigar she found on the back seat of her Rolls that morning, though it was not her brand, and it had been saturated with enough odorless, tasteless TWT to dose

Madison Square Garden. It is of course impossible to ever know exactly what transpired that day in that most sacrosanct and guarded and unpublic of rooms— but we have the text of the press release that ensued, and we do know that all GM products subsequent to 1989 burn alcohol instead of gasoline, and exhibit a sharp upward curve in safety and reliability.

In Chicago Zack and Jill got a prominent and wealthy realtor-developer and all his tame engineers, ecologists, lawyers and other promotion experts in the middle of a public debate over a massive rezoning proposal. There are no more slums in Chicago, and the developer is, of course, its present mayor.

In Cleveland they got a used car salesman, a TV repairman, a plumber, an auto mechanic, and a Doctor of Philosophy in one glorious afternoon.

In New York they got Mtu Zanje, quite by accident. The renegade white led a force of sixteen New Black Panthers in a smash-and-grab raid on the downtown club where Zack and Jill were playing. Mtu Zanje personally took Jill's purse, and smoked a cigar which he found therein on his way back uptown. Zack and Jill never learned of his death or their role in it, but it is doubtful that they would have mourned.

In Boston they concentrated on policemen, as many as they could reach in two mornings and afternoons, and by the time they left that town it was rocking on its metaphorical foundations. Interesting things came boiling up out of the cracks, and most of them have since decomposed in the presence of air and sunlight.

In Portland, Maine, Zack figured a way to plant a timed-release canister in the air-conditioning system of that city's largest Welfare Center. A great many people voluntarily left the welfare roll over the ensuing month, and none have yet returned—or starved. There are, of course, a lot of unemployed caseworkers. . . .

And then they were on their way home to Halifax. But this is a listing only of the headlines that Zack and Jill left behind them—not even of everything that happened on that trip. Not even of everything important; at least, not to Zack and Jill.

In Quebec a laundry van just missed killing them both, then roared away.

In Ottawa they went out for a late night walk just before a tremendous explosion partially destroyed their motel. It had apparently originated in the room next to theirs, which was unoccupied.

In Toronto they were attacked on the streets by what might have been a pair of honest muggers, but by then they were going armed and they got one apiece.

In Detroit the driver of the cab they had taken (at ruinous expense) to eliminate a suspected tail apparently went mad and deliberately jumped a divider into high-speed oncoming traffic. In any car crash, the Law of Chaos prevails, and in this instance it killed the driver and left Zack and Jill bruised and shaken but otherwise unharmed.

They knew enemy action when they saw it, and so they did the most confusing thing they could think of: stopped showing up for their scheduled gigs, but kept on following the itinerary. They also adopted reasonably ingenious disguises and, with some trepidation, stopped traveling together. Apparently the combination worked; they were not molested again until they showed up for the New York gig to break the pattern, and then only by Mtu Zanje, which they agreed was coincidence. But it made them thoughtful, and they rented several hours of complete privacy in a videotape studio before leaving town.

And on the road to Boston they each combed their memory for friends remembered as One Of The Nice Ones, people they could trust, and in that city they met in the Tremont Street Post Office and spent an hour addressing and mailing VidCaset Mailer packs. Each pack contained within it, in addition to its program material, a twenty-second trailer holding five hundred hits of TWT in blotter form—a smuggling innovation of which Zack was sinfully proud.

They had not yet taken TWT themselves, but their decision was made. They agreed at the end of that day to take it together when they got back to Halifax. They would do it in the Scorpio, alone together, in the dressing room where Wesley George had died.

They waited until well after closing, after Finnegan and the Shadow had locked up behind them and driven away the last two cars in the parking lot. Then they waited another hour to be sure.

The night was chill and still, save for the occasional distant street sounds from more active parts of town. There was no moon and the sky was lightly overcast; darkness was total. They waited in the black together, waiting not for any particular event or signal but only until it felt right, and they both knew that time without words. They were more married already than most couples get to be in a lifetime, and they were no longer in any hurry at all.

When it was time they rose from their cramped positions behind the building's trash compactor and walked stealthily around to the front of the building, to the descending stairway that led to the outer door of the dressing room. Like all of Finnegan's regulars they knew how to slip its lock, and did so with minimal noise.

As soon as the door had clicked shut behind them, Jill heaved a great sigh, compounded of relief and fatigue and *déjà vu*. "This is where it all started," she breathed. "The tour is over. Full circle."

Zack looked around at pitch blackness. "From the smell in here, I would guess that it was Starship Earth played here tonight."

Jill giggled. "Still living on soybeans, too. Zack, can we put the light on, do you think?"

"Hmmm. No windows, but this door isn't really tight. I don't think it'd be smart, hon."

"How about a candle?"

"Sold. Let me see—ouch!—if the Starship left the—yeah, here's a couple." He struck a light, and started both candles. The room sprang into being around them, as though painted at once in broad strokes of butter and chocolate. It was, after a solid month of perpetually new surroundings, breathtakingly familiar and comfortable. It lifted their hearts, even though both found their eyes going at once to the spot on which Wesley George had fallen.

"If your ghost is here, Wesley, rest easy, man," Zack said quietly. "It got covered. And we're both back to do truth ourselves. They killed you, man, but they didn't stop you."

After a pause, Jill said, "Thank you, Wesley," just as quietly. Then she turned to Zack. "You know, I don't even feel like we *need* to take the stuff, in a place."

"I know, hon, I know. We've been more and more honest with each other, opened up more every day, like the truth was gonna come sooner or later so we might as well get straight now. I guess I know you better than I've ever known any human, let alone any woman. But if fair is fair and right is right we've *got* to take the stuff. I wouldn't have the balls not to."

"Sure. Come on—Wesley's waiting."

Together they walked hand in hand, past the cigar-burn in the rug, to Wesley's dying place. The whisper of their boots on the rug echoed oddly in the sound-proof room, then faded to silence.

"The door was open that night," Jill whispered.

"Yeah," Zack agreed. He turned the knob, eased the door open and yelped in surprise and fright. A bulky figure sat on the stage ten feet away, half-propped against an amp, ankles crossed before it. It was in deep shadow, but Zack would have known that silhouette in a coal cellar. He pushed the door open wider, and the candlelight fell on the figure, confirming his guess.

"Finnegan!" he cried in relief and astonishment. "Jesus Christ, man, you scared me. I swear I saw you leave an hour ago."

"Nope," said the barkeep. He was of medium height and stocky, bald as a grape but with fuzzy brown hair all over his face and neck. It was the kind of face within which the unbroken nose was incongruous. He scratched his crinkly chin with a left hand multiply callused from twenty years of guitar and dobro and mandolin and fiddle, and grinned what his dentist referred to as the Thousand Dollar Grin. "You just thought you did."

"Well shit, yeah, so it seems. Look, we're just sort of into a little head thing here if that's cool, meant to tell you later . . ."

"Sure."

A noise came from behind Zack, and he turned quickly to Jill. "Look, baby, it's Finn—"

Jill had not made the noise, nor did she make one now. Sziller had made the noise as he slipped the lock on the outside door, and he made another one as he snapped the hammer back on the silenced Colt. It echoed in the dressing room. Zack spun back to Finnegan, and the barkeep's right hand was up out of his lap now and there was a .357 Magnum in it.

Too tired, Zack thought wearily, *too frigging tired. I wasn't cautious enough and so it ends here.*

"I'm sorry, Jill," he said aloud, still facing Finnegan.

"I," Finnegan said clearly and precisely, "am a bi-federal agent, authorized to act in either the American or the Canadian sector. Narcotics has been my main turf for years now."

"Sure," Zack agreed. "What better cover for a narc than a musician?"

"This one," Finnegan said complacently. "I always hated being on the road. Halifax has always been a smuggler's port—why not just sit here and let the stuff come to me? All the beer I can drink—"

Sziller was going through the knapsack Jill had left

ьy the door, without taking his eyes or his gun off them for an instant.

"So how come you're in bed with Sziller?" Zack demanded. Sziller looked up and grinned, arraying his massive beard like a peacock's tail.

"George blew my cover," Finnegan said cheerfully. "He knew me from back when and spilled the soybeans. If he'd known you two were regulars here he'd likely have warned you. So after Sziller did him in and then . . . found out he had not adequately secured the goods . . . he naturally came straight to me."

"Finnegan's got a better organization than we do," Sziller chuckled. His voice was like a lizard's would sound if lizards could talk. "More manpower, more resources, more protection."

"And Sziller knew that TWT would mean the end of me too if it got out. He figured that our interests coincided for once—in a world of truth, what use is a narc? How can he work?"

Much too goddam tired, Zack told himself. *I'm hallucinating.* Finnegan appeared to be winking at him. Zack glanced to see if Jill were reacting to it, but her eyes were locked on Sziller, whose eyes were locked on her. Zack glanced swiftly back, and Finnegan still appeared to be winking, and now he was waving Zack toward him. Zack stood still; he preferred to die in the dressing room.

"He took a gamble," Finnegan went on, "a gamble that I would go just as far as he would to see that drug destroyed. Well, we missed you in Quebec and Ottawa and Toronto, and you fooled us when you went to Portland instead of your gig in Bangor, but I guess we've got you now."

"You're wrong," Jill said, turning to glare at Finnegan. "It's too late. You're both too late. You can kill us, but you can never recall the truth now."

"People forget headlines," Sziller sneered confidently. "Even a month of headlines. Nothing."

"You're still wrong," Zack said, staring in confusion

from Sziller to Jill to the gesticulating Finnegan. "Wε
put about thirty tapes and TWT samples in the mail—"

"Jerks," Sziller said, shaking his head. "Outthought
every step of the way. Look, sonny, if you want to
move a lot of dope with minimum risk, where do you
get a job?" He paused and grinned again. "The Post
Office, dummy."

"No," Zack and Jill said together, and Finnegan
barked "*Yes*" quite sharply. They both turned to look at
him.

"You can bug any room with a window in it, children," he said wearily. "And that dressing room, of
course, has always been bugged. Oh, *look*, dammit."

He held up a VidCaset Mailer pack with broken
seals, and at *last* they both started forward involuntarily toward it, and as he cleared the dressing room
doorway Zack finally caught on, and he reached behind him and an incredible thing happened.

It must be borne in mind that both Zack and Jill
had, as they had earlier recognized, been steadily raising the truth level between them for over a month,
unconsciously attempting to soften the blow of their
first TWT experience. The Tennessee preacher earlier
noted had once said publicly that all people are born
potentially telepathic—but that if we're ever going to
get any message-traffic capacity, we must first shovel
the shit out of the Communications Room. This room,
he said, was called by some the subconscious mind.
Zack and Jill had almost certainly been exposed to at
least threshold contamination with TWT, and they
were, as it happens, the first subjects to be a couple
and very much in love. They had lived together
through a month that could have killed them at any
time, and they were already beginning to display mi-
nor telepathic rapport.

Whatever the reasons, for one fractionated instant
their hands touched, glancingly, and Jill—who had
seen none of Finnegan's winking and almost nothing
of his urgent gestures—knew all at once exactly what

I have done things that horrify me, things that diminish me, but I did good things, too, and I have been striving every minute toward a world in which my job didn't exist, in which nobody had to shoulder that load. I've been working to put myself out of a job, without the faintest shred of hope, for over ten years—and now it's Christmas and I'm free, I'm fucking *FREE*. That makes me so happy I could go down to the cemetery and dig up Wes and kiss him on the moldy lips, so happy I'll feel just *terrible* if I can't talk you two out of killing me.

"My job is finished, now—nobody knows it but you and me, but it's all over but the shouting. And in gratitude to you and Wes I intend to use my last gasp of power and influence to try and keep you two alive when the shit hits the fan."

"Huh?"

"I kind of liked your idea, so I let your VidCaset packs go through. But first I erased 'em and rerecorded. Audio only, voice out of a voder, nothing identifying you two. That won't fool a computer for long, they're all friends of yours, but it buys us time."

"For what?" Jill asked.

"Time to get you two underground, of course. How would you like to be, oh, say, a writer and her husband in Colorado for six months or so? You'd look good as a blonde."

"Finnegan," Zack said with great weariness, "this all has a certain compelling inner consistency to it, but you surely understand our position. Unless you can prove any of this, we're going to have to shoot it out."

"Why you damned fools," Finnegan blazed, "what're you wasting time for? You've got some of the stuff with you—*give me a taste.*"

There was a pause while the pair thought that over. "How do we do this?" Jill asked at last.

"Put your guns on me," Finnegan said.

They stared.

"Come on, dammit. For now that's the only way

we can trust each other. Just like the world out there—
guns at each other's heads because we fear lies and
treachery, the sneak attack. Put your fucking guns on
me, and in an hour that world will be on its way out.
Come on!" he roared.

Hesitantly, the two brought up their guns, until all
three weapons threatened life. Jill's other hand
brought a tiny stoppered vial from her pants. Slowly,
carefully, she advanced toward Finnegan, holding out
the truth, and when she was three feet away she saw
Finnegan grin and heard Zack chuckle, and then she
was giggling helplessly at the thought of three solemn
faces above pistol sights, and all at once all three of
them were convulsed with great racking whoops of
laughter at themselves, and they threw away their
guns as one. They held their sides and roared and
roared with laughter until all three had fallen to the
floor, and then they pounded weakly on the floor and
laughed some more.

There was a pause for panting and catching of
breath and a few tapering giggles, and then Jill un-
stoppered the vial and upended it against each prof-
fered fingertip and her own. Each licked their finger
eagerly, and from about that time on everything began to be all right. Literally.

An ending is the beginning of something, always.

TIDBIT: a triple Feghoot and a cartoon

NOTE: the character of Ferdinand Feghoot, punster extraordinary, space and time traveler, and charter member of the Society for the Aesthetic Rearrangement of History, was created by Grendel Briarton, and is used here with his permission.

Three-Time Winner

Even Ferdinand Feghoot could be outpunned on occasion—but he always rose to the challenge.

There was, for instance, the time he conducted a crew of new S.A.R.H. recruits—all from late twentieth-century Terra—on a training study of Carter's World, a newly established agricultural colony attempting to support itself by the export of edible nuts. Barely into their second generation, and having yet to show a profit, the colonists were technologically backward—yet they showed a surprising ingenuity in the use of their few advantages. It was this resourcefulness that Feghoot was demonstrating to his rookies.

"Look at the perfection with which these streets are graded," exclaimed one student. "Earth-moving machinery on this scale is strictly high technology stuff. How can they do it?"

"A new alleyway is being constructed nearby," said Feghoot. "Let us walk that way while I explain." As they strolled, he told his students that countless centuries before, the Carter's World system had been inhabited by a now-vanished race of giants. This very planet had served them for a nursery, and among the many artifacts they had left were thousands of children's blocks, immense and precision-cut. "You simply jack one up onto logs, bring it where you want it, put collapsible jacks underneath, snake out the logs, spread soil more or less evenly beneath, and collapse the jacks."

"I see," said the student. "It's not graded road at all; it's a simple hammered-earth base."

"That's right," Feghoot went on smoothly. "You just hit the road, jack, and don't come back no mo'."

His students registered dismay and anguish.

"Isn't that right, old-timer?" Feghoot demanded of an ancient Carterian standing by the mouth of the newly completed alley they had just reached.

"Ah'm afraid not, suh," said the senior citizen, and the students giggled at Feghoot's discomfiture. "Oh, we used to do it that way, but it was far too much trouble. It's the soil heah, you see: the very same soil which produced our famous cashews is so high in clay content that a child could roll out a road of it. Then we simply use a system of lenses to bake it into hardness. Ah've just completed this alley mahself, and ah'm just a retired professor of Sports History, much too old and feeble to handle hydraulic jacks.

"So you see," he finished, eyes twinkling, "Mah hammered alley is really cashew's clay."

Howls of agony rose from the students, but Feghoot never hesitated. "And he," he said, turning to his students, "is clearly the gradist."

"Sorry, Mr. Griffin: he says he can't see you now."

7
APOGEE

He sat on plush leather in the finest, most opulent office in town, surveying a desk on which even a careless pilot could have landed a helicopter. Flicking an entirely imaginary speck of lint from the lapel of his newest four-hundred-dollar suit, he yawned for perhaps the twentieth time since his secretaries had gone home for the day, and stifled the yawn with an exquisitely manicured hand. His countenance was that of a man with perfect health, job security, much money, and considerable prestige—with a paradoxical frown overlaid.

"Hell," he said succinctly and most uncharacteristically.

"Yeah?" said the demon which appeared flaming beyond the desk.

The temperature in the room rose sharply, but the seated man did not (as a matter of fact, could not) sweat. He squinted at the blazing horned creature and automatically moved his Moroccan leather cigar box away from it. "You want to tone that down a bit?" he said, scowling.

"Listen," it told him, "with the price of a watt these days, you should turn out the lights and put a mirror behind me." But its fiery brilliance moderated to a cheery glow, and the carpet stopped smelling bad. It sat down on thin air, tail coiled, and blew a perfect smoke ring. "Now, what's on your mind?"

He hesitated; took the plunge. "I'm not satisfied."

The demon sneered. "A beef, huh? You guys gimme a pain. You want the Moon for a soul like yours?"

"Now wait a minute," he said indignantly, with just a touch of fear. "We've got a contract."

"Yeah, yeah," it sighed. "And you want to talk fine print. You guys read too many stories. All right, let's haul out the contract and get this over with."

A large piece of foolscap appeared between them on the desk, smouldering around the edges. It was covered with minuscule type, and one of the signatures glistened red.

"Standard issue contract, with bonus provisions contingent on your promise to deliver a large consignment of souls other than your own, as described in appended schedule A-2 . . ." The appendix materialized beside the contract, and the demon looked it over. "Seems to be in order. What's the beef?"

"I'm not satisfied," he repeated, and glared uncomfortably at the demon.

"Oh for cryin' out loud," it burst out, "what do you want from my life? You got everything you asked for. I honor my service contracts, I supplied everything requested and I mean everything. I *worked* for you, baby."

"I don't care," he said petulantly. "I'm not happy. It's right there in the appendix, the Lifetime Approval Option. I've got to *enjoy* all that you give me. And I don't."

"Look," the demon said angrily, "I did my best for you, pal—you've got all I can give you. Unbelievable riches, total health, raw power, the job you always wanted and complete autonomy. You can say any dumb thing that comes into your head—and believe me, you've said some lulus—and people agree with you. You can make the wildest bonehead decisions and they work out okay. You couldn't louse up if you tried, and believe you me it's taken some doing. So what's not to enjoy?"

He glared at it, his jowls quivering. "I'm bored, dammit. There's nothing left to achieve."

"It's your own fault," said the demon. "You insisted on having everything right away, and so you ran out of dreams too fast." It sneered at him. "Greedy."

"I don't care," he snapped. "You made a deal and I want satisfaction. Literally."

The demon stood and began pacing the floor, trailing wisps of blue smoke. "Look," it said irritably, "there's nothing more I can *do*. You've got the whole works."

"It's not enough. I'm bored."

The demon looked harassed, then thoughtful. "Maybe there's a way," it said slowly.

"Yes," he prodded eagerly.

"It's a way-out idea, but it just might work. The only thing you haven't tried. I'll turn you into a woman, and . . ."

"No," he said firmly.

It grimaced. "Worth a try. Well, I guess there's only one possibility, then."

"Well, come on, come on. Out with it."

"I'll turn you into a masochist, and let the whole job come down around your ears." The demon smiled. "Take a big bite out of my work load."

"Are you out of your mind?" he exploded.

"Think about it," it said reasonably. "There's nowhere to go from here but back downhill, and you could enjoy that just as much as the ride up. Don't you understand? You'd be a *masochist*. You'll lose everything I've ever given you with just as much joy as you experienced in receiving it, only this time you'll be *doing it all yourself*, through your own natural ineptitude. All I'll do is help you appreciate it."

He started to say that it was the craziest thing he'd ever heard, and paused. He was silent for a long time, rubbing his five o'clock shadow, and the demon waited. At last he cleared his throat.

"Do you really think it's feasible?" he asked.

"*Thought* so," said the demon with sly satisfaction. "You've been kidding yourself all these years; this is what you really wanted all along." He began an angry retort, but paused. All at once he experienced a flash of nostalgia for his ulcer. It *might* be nice to whimper again . . .

"All right," he said suddenly. "Do it."

"It's done."

The demon disappeared, leaving behind it the traditional smell of brimstone (with added petroleum derivatives) and a scorched carpet.

He discovered that his feet hurt, and realized with what was now the closest thing to glee that he could experience that he was sweating profusely. The demon was right—*this* was what he had really craved all along, this was what he had been born for. The fall would be more spectacular than the rise. His head began to ache dully.

Picking up the special phone, he made two calls, then dialed his unlisted home number. "Hello, Pat? Dick. Sorry I'm late, dear. I'll be sleeping here tonight. I have to meet early tomorrow with Ron and Gordon about some plumbers. Yes, I'll see you tomorrow night. What? No, dear, nothing's wrong. Everything is fine. Everything is just fine. Good night, dear."

He hung up and looked across the room at the presidential seal over the door. He began to laugh, and then he cried, and continued to cry for months thereafter.

TIDBIT: two puns

A Standing Joke (reprinted from
Callahan's Crosstime Saloon)
In the year 2744 a human survey team discovered a
planet whose sole inhabitant was an enormous humanoid, three miles high and made of something very
like granite. At first it was mistaken for an immense
statue left by some vanished race of giants, for it
squatted motionless on a vast rocky plain, exhibiting
no outward sign of life. It had legs (two), but apparently never rose to walk on them. It had a mouth, but
never ate or spoke. It had what appeared to be a perfectly functional brain, the size of a fifty-story con-
dominium, but the organ lay dormant, electrochemi-
cal activity at a standstill. Yet it lived.

This puzzled hell out of the scientists, who tried everything they could think of to elicit some sign of life
from the behemoth—in vain. It just squatted, motionless and seemingly thoughtless, until one day a xenobiologist, frustrated beyond endurance, screamed,
"How could evolution give legs, mouth, and brain to a
creature that doesn't *use* them?"

It happened that he was the first one to ask a direct
question in the thing's presence. It rose with a thun-
derous rumble to its full height, scattering the clouds,
pondered for a second, boomed, "IT COULDN'T,"
and squatted down again.

"Migod," exclaimed the xenobiologist. "Of course! *It
only stands to reason.*"

The Snoopy Scientist

Two Egyptologists, Freenbean and Fonebone, were accustomed to correspond with each other by means of pictograms of their own devising. One day Fonebone received an urgent letter from his old friend. Freenbean, it seemed, had accidentally defiled a sarcophagus, *before* deciphering the curse which was inscribed thereon. It translated as something like, "May you fall into the latrine just as a platoon of Sumerians is finishing a prune stew and seven barrels of beer." Clearly shaken, Freenbean ended his letter with the following hieroglyph:

!¢&!$#@!%!!

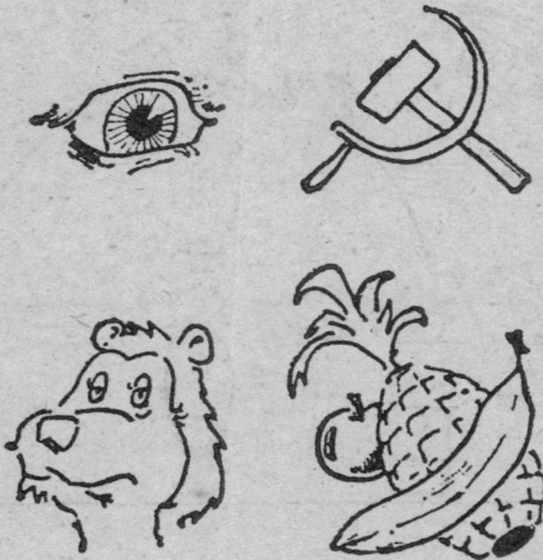

which, of course, translates: "WILL THE CURSE / I READ / BEAR FRUIT?"

Fonebone's reassuring reply was immediate:

8
NO RENEWAL

<hr style="height:3px; border:none; background:repeating-linear-gradient(90deg,#000 0,#000 2px,#fff 2px,#fff 4px)">

Douglas Bent Jr. sits in his kitchen, waiting for his tea to heat. It is May 12, his birthday, and he has prepared wintergreen tea. Douglas allows himself this estravagance because he knows he will receive no birthday present from anyone but himself. By a trick of Time and timing, he has outlived all his friends, all his relatives. The concept of neighborliness, too, has predeceased him; not because he has none, but because he has too many.

His may be, for all he knows, the last small farm in Nova Scotia, and it is bordered on three sides by vast mined-out clay pits, gaping concentric cavities whose insides were scraped out and eaten long ago, their husk thrown away to rot. On the remaining perimeter is an apartment-hive, packed with antlike swarms of people. Douglas knows none of them as individuals; at times, he doubts the trick is possible.

Once Douglas's family owned hundreds of acres along what was then called simply the Shore Road; once the Bent spread ran from the Bay of Fundy itself back over the peak of the great North Mountain, included a sawmill, rushing streams, hundreds of thousands of trees, and acre after acre of pasture and hay and rich farmland; once the Bents were one of the best-known families from Annapolis Royal to Bridgetown, their livestock the envy of the entire Annapolis Valley.

Then the petrochemical industry died of thirst. With it, of course, went the plastics industry. Clay suddenly became an essential substitute—and the Annapolis Valley is mostly clay.

Now the Shore Road is the Fundy Trail, six lanes of high-speed traffic; the Bent spread is fourteen acres on the most inaccessible part of the Mountain; the sawmill has been replaced by the industrial park that ate the clay; the pasture and the streams and the farmland have been disemboweled or paved over; all the Bents save Douglas Jr. are dead or moved to the cities; and no one now living in the Valley has ever seen a live cow, pig, duck, goat or chicken, let alone envied them. Agribusiness has destroyed agriculture, and synthoprotein feeds (some of) the world. Douglas grows only what crops replenish themselves, feeds only himself.

He sits waiting for the water to boil, curses for the millionth time the solar-powered electric stove that supplanted the family's woodburner when firewood became impossible to obtain. Electric stoves take too long to heat, call for no tending, perform their task with impersonal callousness. They do not warm a room.

Douglas's gnarled fingers idly sort through the wintergreen he picked this morning, spurn the jar of sugar that stands nearby. All his life Douglas has made wintergreen tea from fresh maple sap, which requires no sweetening. But this spring he journeyed with drill and hammer and tap and bucket to his only remaining maple tree, and found it dead. He has bought maple-flavored sugar for this birthday tea, but he knows it will not be the same. Then again, next spring he may find no wintergreen.

So *many* old familiar friends have failed to reappear in their season lately—the deer moss has gone wherever the hell the deer went to, crows no longer raid the compost heap, even the lupens have decreased in number and in brilliance. The soil, perhaps made self-conscious by its conspicuous isolation, no longer bursts with life.

Douglas realizes that his own sap no longer runs in the spring, that the walls of his house ring with no voice save his own. If a farm surrounded by wasteland cannot survive, how then shall a man? *It is my birthday*, he thinks, *how old am I today?*

He cannot remember.

He looks up at the goddamelectricclock (the family's two-hundred-year-old cuckoo clock, being wood, did not survive the Panic Winter of '94), reads the date from its face (there are no longer trees to spare for fripperies like paper calendars), sits back with a grunt. *2049, like I thought, but when was I born?*

So many things have changed in Douglas's lifetime, so many of Life's familiar immutable aspects gone forever. The Danielses to the east died childless: their land now holds a sewage treatment plant. On the west the creeping border of Annapolis Royal has eaten the land up, excreting concrete and steel and far too many people as it went. Annapolis is now as choked as New York City was in Douglas's father's day. Economic helplessness has driven Douglas back up the North Mountain, step by inexorable step, and the profits (he winces at the word) that he reaped from selling off his land parcel by parcel (as, in his youth, he bought it from his ancestors) have been eaten away by the rising cost of living. Here, on his last fourteen acres, in the two-story house he built with his own hands and by Jesus *wood*, Douglas Bent Jr. has made his last stand.

He questions his body as his father taught him to do, is told in reply that he has at least ten or twenty more years of life left. *How old am I?* he wonders again, *forty-five? Fifty? More?* He has simply lost track, for the years do not mean what they did. It matters little; though he may have vitality for twenty years more, he has money for no more than five. Less, if the new tax laws penalizing old age are pushed through in Halifax.

The water has begun to boil. Douglas places wintergreen and sugar in the earthenware mug his mother made (back when clay was dug out of the backyard with a shovel), moves the pot from the stove, and pours. His nostrils test the aroma: to his dismay, the fake smells genuine. Sighing from his belly, he moves to the rocking chair by the kitchen window, places the mug on the sill, and sits down to watch another sunset. From here Douglas can see the Bay, when the wind is right and the smoke from the industrial park does not come between. Even then he can no longer see the far shores of New Brunswick, for the air is thicker than when Douglas was a child.

The goddamclock hums, the mug steams. The winds are from the north—a cold night is coming, and tomorrow may be one of the improbable "bay-steamer" days with which Nova Scotia salts its spring. It does not matter to Douglas: his solar heating is far too efficient. His gaze wanders down the access road which leads to the highway; it curves downhill and left and disappears behind the birch and alders and pine that line it for a half mile from the house. If Douglas looks at the road right, he can sometimes convince himself that around the bend are not strip-mining shells and brick apartment-hives but arable land, waving grain and the world he once knew. Fields and yaller dogs and grazing goats and spring mud and tractors and barns and goat berries like stockpiles of B-B shot . . .

Douglas's mind wanders a lot these days. It has been a long time since he enjoyed thinking, and so he has lost the habit. It has been a long time since he had anyone with whom to share his thoughts, and so he has lost the inclination. It has been a long time since he understood the world well enough to think about it, and so he has lost the ability.

Douglas sits and rocks and sips his tea, spilling it down the front of his beard and failing to notice. *How old am I?* he thinks for the third time, and summons enough will to try and find out. Rising from the rocker with an effort, he walks on weary wiry legs to the living room, climbs the stairs to the attic, pausing halfway to rest.

My father was sixty-one he recalls as he sits, wheezing, on the stair *when he accepted euthanasia. Surely I'm not that old. What keeps me alive?*

He has no answer.

When he reaches the attic, Douglas spends fifteen minutes in locating the ancient trunk in which Bent family records are kept. They are minutes well spent: Douglas is cheered by many of the antiques he must shift to get at the trunk. Here is the potter's wheel his mother worked; there the head of the axe with which he once took off his right big toe; over in the corner a battered peavey from the long-gone sawmill days. They remind him of a childhood when life still made sense, and bring a smile to his grizzled features. It does not stay long.

Opening the trunk presents difficulties—it is locked, and Douglas cannot remember where he put the key. He has not seen it for many years, or the trunk for that matter. Finally he gives up, smashes the old lock with the peavey, and levers up the lid (the Bents have always learned leverage as they got old, working efficiently long after strength has gone). It opens with a shriek, hinges protesting their shattered sleep.

The past leaps out at him like the woes of the world from Pandora's Box. On top of the pile is a picture of Douglas's parents, Douglas Sr. and Sarah, smiling on their wedding day, Grandfather Lester behind them near an enormous barn, grazing cattle visible in the background.

Beneath the picture he finds a collection of receipts for paid grain bills, remembers the days when food was cheap enough to feed animals, and there were animals to be fed. Digging deeper, he comes across canceled checks, insurance policies, tax records, a collection of report cards and letters wrapped in ribbon. Douglas pulls up short at the hand-made rosary he gave his mother for her fifteenth anniversary, and wonders if either of them still believed in God even then. Again, it is hard to remember.

At last he locates his birth certificate. He stands, groaning with the ache in his calves and knees, and threads his way through the crowded attic to the west window, where the light from the setting sun is sufficient to read the fading document. He seats himself on the shell of a television that has not worked since he was a boy, holds the paper close to his face and squints.

"May 12, 1989," reads the date at the top.

Why, I'm sixty years old he tells himself in wonderment. *Sixty. I'll be damned.*

There is something about that number that rings a bell in Douglas's tired old mind, something he can't quite recall about what it means to be sixty years old. He squints at the birth certificate again.

And there on the last line, he sees it, sees what he had almost forgotten, and realizes that he was wrong— he will be getting a birthday present today after all.

For the bottom line of his birth certificate says, simply and blessedly, ". . . expiry date: May 12, 2049."

Downstairs, for the first time in years, there is a knock at the door.

Afterword to "No Renewal"

I wrote "No Renewal" in the belief that the oil crisis (a) existed and (b) might just chew up my adopted homeland. If you got no petroleum, you got no plastic; if you got no plastic, you need a lot of clay; and the Annapolis Valley is some of the finest farmland on earth spread over one of the biggest clay deposits in North America.

Happily, since the story was written I read that a feller has perfected a commercially viable method of making plastic from corn (situation comedies do it all the time), relieving some of the immediate threat.

But I stand by the story. We could yet have a corn crisis . . .

Douglas Bent, Jr. is imaginary, as are his parents. But his grandparents Lester and Beth are real: they were my neighbors for three years on the Fundy Shore. My second-nearest neighbors—darn near within eyeshot from the rooftree. Lester's parents, Don and Achsa, live next door to him. Don is . . . well, I've never asked, but I'd say he'll never see seventy again, and I've seen him chainsaw logs in his bedroom slippers. He says steel-toed boots "dull too many blades." He can look out the window at the color of the Bay and give you an accurate weather forecast for the next week (which God himself couldn't do in Nova Scotia). Once my pal Charlie asked him straight out if there was any truth to the rumor that advancing age diminished one's carnal appetites. "Oh, sure, that's true," Don said at once. "You first notice it come on about ten minutes after they lay you in the ground."

This salty soul is precisely what is lacking in my imaginary Douglas Bent, Jr. It is there, now, in Don, and in Achsa, and in Lester and Beth and, apparently, in all of their kids. It is there, now, in all of the tough, smart breed of farmers to which the Bents belong, and I mean no insult to them or their descendants.

But there are fewer of them on the Mountain than there used to be.

I believe there is a . . . a force or a thing abroad on the land that is eating away the strong backbones of our farmers. It may be "only" as subtle a thing as a climate of opinion; it may well be, as some claim, deliberate economic conspiracy. Irrelevant: we cannot spare those farmers. Every time one of the farms in the Valley goes back to scrub and alders, a generation of energy and sweat is thrown away forever.

And it's partly your fault. Do you know what the policies of *your* government are with regard to agriculture? Or did you think food grows on supermarket shelves? And do you really believe that agribusiness will continue to feed you once it's got you over a barrel, when it sees a chance to efficiently "centralize" the profits with, say, a corn crisis . . . ?

Through My Eyes
Meaning no disrespect to the many fine artists (including Vince Di Fate and Jack Gaughan) who have depicted this character elsewhere, here is my own interpretation of Mike Callahan, proprietor of *Callahan's Crosstime Saloon*. If you happen to have different preconceptions, feel free to keep 'em.

Mike Callahan

Silly Weapons Throughout History

People keep sending me their fanzines—amateur publications concerning sf and related subjects, and spanning the spectrum from mimeographs to four color offset. As with amateur efforts of any kind, some are just dreadful and some are sublime. One of the most piquant I have seen is a little 'zine out of Florida called *The Tabebuian.* It is the size of a pocket-calculator instruction pamphlet, much better printed, published by Mensa members Dave and Mardi Jenrette. I can attest to the fact that David's sense of humor is D. Jenrette. I wrote him a letter asking why, if Mensa people were so smart, they had named their organization after the Latin word for table (*mensa*) rather than mind (*mens,* an early example of unconscious sexism). He replied that the club's name is in fact derived from *menses*, and refers to their periodic meetings. (I gave this riposte a standing ovulation.)

Anyroad, one of the *Tab*'s running departments for a while was a feature called "Silly Weapons Throughout History." The first one I saw was the Jell-O Sword, a short-lived weapon rendered obsolete by the subsequent invention (a week later) of the bronze sword. Inspired, I retired to my Fantasy Workbench, and over the next few days I hammered out the following Silly Weapons:

The Swordbroad: Invented by a tribe of fanatical male chauvinists, the Prix, this armament consisted of a wife gripped by the ankles and whirled like a flail (Prix warriors made frequent jocular allusion to the sharp cutting edge of their wives' tongues). The weapon died out, along with the Prix, in a single generation—for tolerably obvious reasons.

The Rotator: A handgun in which the bullets are designed to rotate as well as revolve, presenting an approximately even chance of suicide with each use.

The Bullista: A weapon of admittedly limited range which attempted to sow confusion among the enemy by firing live cows into their midst, placing them upon a dilemma of the horns. (Also called the Cattling Gun.)

The Arbalust: A modification of the bullista, which sought to demoralize and distract the enemy by peppering their encampments with pornographic pictures and literature—yet another dilemma of the horns.

The Dogapult: Another modification of the bullista; self-explanatory.

The Cross 'Bo: Yet another modification of the bullista, this weapon delivered a payload of enraged hobos. Thus gunnery officers had a choice between teats, tits, mastiffs or bindlestiffs.

The Blunderbus: A hunter-seeker weapon which destroys the steering box in surface mass transit.

The Guided Missal: Originally developed as a specific deterrent to the Arbalust; as, however, it is hellishly more destructive, its use is now restricted by international convention to Sundays.

The Slingshit: self explanatory; still used in politics and in fandom.

And, of course, such obvious losers as the *foot ax, relish gas, studded mice,* and the effective but disgusting *snotgun.*

Ironically enough, since I wrote the above I have learned that the United States has recently been bombed several times by commercial airliners. Honest to God. True fact, documentation available. Airliner toilet holding tanks often leak, resulting in accumulations of blue ice on the fuselage during high-altitude flight. The blue ice is composed of roughly equal parts of urine, feces and blue liquid disinfectant. If the plane is required to make its landing descent rapidly enough, chunks of blue ice ranging to upwards of two hundred pounds can—and *do*—break loose and shell the countryside. I have seen a photograph of a roofless, floorless apartment that was demolished by a one-hundred-and-fifty-pound chunk of Blue Ice. It pulped an electric range in the apartment below. All the occupants escaped unscathed, but considerably unnerved.*

Now if *that* ain't a silly weapon, I don't know what is.

So it doesn't matter if you were cautious enough not to make your home near any strategic military targets. If you live anywhere near a commercial airport, you stand a chance of being attacked by an Icy B.M.

---

*Science fiction devotees beware: it's said to be exceptionally terrible if that stuff hits a Fan . . .

9
OVERDOSE

Moonlight shattered on the leaves overhead and lay in shards on the ground. The night whispered dementedly to itself, like a Zappa minuet for paintbrush and tea kettle, and in the distance a toad farted ominously.

I was really stoned.

I'd never have gotten stoned on sentry duty in a real war, but there hadn't been much real fighting to speak of lately (this was just before we got out), and you have to pass the time somehow. And it just so happened that as I was getting ready to leave for the bush, a circle of the boys was Shotgunning.

Shotgunning? Oh, we do a lot of that. It works like so: the C.O. (. . . . "or whomever he shall appoint . . .") fills a pipe from the platoon duffel bag, fires it up, takes a few hits to get it established, and then breaks open a shotgun and inserts the pipe in one of the barrels. He raises it to his lips and blows a mighty blast down the bore, and someone on the other end takes an *enormous* hit from the barrel.

The C.O. then passes the Shotgun . . .

So, as I say, I was more ruined than somewhat as I contemplated the jungle and waited for my relief. Relief? Say, you can take your meditation and your yoga and your za-zen—there's nothing on earth for straightening your head like a night in the jungles of Vietnam. Such calm, such peace, such utter tranquility.

Something crackled in the bush behind me, and my M-32 went off with a Gotterdammerung crash two inches from my left ear. As I whirled desperately about, Corporal Zeke Busby, acting C.O. and speed-freak extraordinaire, levitated a graceful foot above the surrounding vegetation and came down rapping.

"Yas indeed, private, yas indeed alert and conscientious as ever yas and a good thing, too, a good thing but if I may make so bold and without wishing to appear unduly censorious, would you for Chrissake point that fuckin' thing somewhere else?" Corporal Zeke had once been a friend of Neal Cassidy's, for perhaps just a bit too long.

"Sure thing, Corp," I mumbled, shifting the rifle. My eardrum felt like Keith Moon's tom-tom.

"Yas and a signal honor, a signal honor my man your gratitude will no doubt be quite touching but I assure you before you protest that I consider you utterly worthy worthy worthy to the tips of your boogety-boogety shoes."

A signal honor? He could only mean . . .

"I have selected you from a field of a dozen aspirants to make the run to Saigon and cop the Platoon Pound."

I was overwhelmed. The last man so honored (a guy named Milligram Mulligan) had burned us for two bricks of Vietnamese cowshit and split for the States—this was indeed a mark of great trust. I tried to stammer my thanks, but Corporal Zeke was off again. ". . . situation of course most serious and grave without at the same time being in any sense of the word *heavy* as I'm sure you dig considering the ramifications of the logistical picture and the inherently inescapable discombobulation manifest in the necessary . . . what I mean . . . that is to say, we've only got fifty bucks to work with." His left eye began to tic perceptibly, almost semaphorically.

"No problem, Corporal Zeke. I've seen action before." Fifty was not much for a platoon, even at Vietnamese prices, but the solution was simple enough—rip off a Gook. "What did you have eyes to score?"

"Yas well based on past performance and an extrapolated estimate of required added increment to offset inflation which some of these lousy bastards they smoke 'til their noses bleed, it seems that something on the close order of five bricks would not be inordinate."

I nodded. "You're faded, Corp. Get me a relief and I'll crank right now." He didn't hear me; he was totally engrossed in his left foot, crooning to it softly. I put the M-32 near him gently and split. When the Old Man says "Cop!", you cop, and ask how soon on the way back.

Deep in the jungle something stirred. Trees moved ungraciously aside; wildlife changed neighborhoods. A space was cleared. In this clearing grew a shimmering ball of force, a throbbing nexus of molecular disruption. It reached a diameter of some thirty feet, absorbing all that it touched, and then stopped growing abruptly. It turned a pale green, flared briefly, and stabilized, emitting a noise like a short in a fifty megavolt circuit—

With something analogous to a gasp, Yteic-Os the Voracious materialized within the sphere, and fell with a horrendous crash to the jungle floor a foot below. Heshe winced—well, not exactly—and momentarily lost conscious control of the pale green bubble, which snapped out of existence at once.

Yteic-Os roared hisher fury (although there was nothing a human would have recognized as sound) and tried to block the green sphere's dissolution by a means indescribable in human speech, something like sticking one's foot in a slamming door. It worked just about as well; the Voracious One nearly lost a pseudopod for hisher trouble.

This was serious.

Yteic-Os was ridiculously ancient—heshe had been repairing hisher third sun on the day when fire was discovered on earth. Entropy is, however, the same for everybody. Yteic-Os had long since passed over into catabolism: hisher energy reserves dwindled by the decade.

This jumping in and out of gravity wells was a hell-ishly exhausting business; for centuries Yteic-Os had sidestepped the problem by using the tame space-warp over which heshe had so laboriously established control. Now the warp was gone, galaxies away by this time, and Yteic-Os had grave doubts as to hisher ability to jump free unassisted.

This world would simply have to serve. Somewhere on this planet must exist a life-form of sufficient vitality to fill Yteic-Os's reserve cells with The Force, and heshe was not called The Voracious for nothing. Heshe extended pseudopods gingerly, questing for data on cerebration-levels, indices of disjunctive thought and the like. Insignificant but potentially useful data such as atmosphere-mix, temperature, radiation-levels and gravity were meanwhile being absorbed below the conscious level by the sensor-modules which studded Yteic-O's epidermis (giving himher, incidentally, the external appearance of a slightly underdone poached egg with pimples).

A pseudopod like a mutant hotdog twitched, began to quiver. Yteic-Os integrated all available data and decided ocular vision was called for. Hastily heshe grew an eye, or something very like one, and looked in the direction pointed by the trembling pseudopod.

Yes, no doubt of it, a sentient life-form, just brim-ming with The Force! Yteic-Os sent a guarded probe, yelped with joy (well, not precisely) as heshe learned that this planet was crawling with sentient beings. What a bountiful harvest!

Yteic-Os cannily withdrew without the other so much as suspecting hisher existence, and began pa-tiently constructing hisher attack.

Well, the plan was simplicity itself: meet Phuc My in a bar, demand to see the goods before paying, pull my gun and depart with the bag. Instead, I left without my pants. How the hell was I supposed to know the bartender had me covered?

So there I crouched, flat broke and *sans culottes*, between two G.I.-cans of reeking refuse in a honky-tonk alleyway, strung out and dodging The Man. It made me homesick for Brooklyn. At least the problem was clear-cut: all I had to do was scare up a pair of pants, five bricks of acceptable smoke, a hot meal and transportation back to my outfit before dawn. Any longer and Corporal Zeke would assume I had burned him, at which point, Temporary Cease-Fire or not, Southeast Asia would become decidedly too warm for me to inhabit. I was not prepared to emulate Milligram Mulligan—ocean-going desertion requires special preparation and a certain minimum of cash, and I had neither.

The possibilities were, as I saw it, dismal. I couldn't rip off a pedestrian without at least a token weapon, and I was morally certain the two garbage cans contained nothing more lethal than free hydrogen sulfide. I couldn't burgle a house without more of the above-mentioned preparation, and I couldn't even borrow money without a pair of pants.

I sure wished I had a pair of pants.

A giggle rippled down the alleyway, and I felt my spine turn into a tube of ice-cold Jell-O. I peered over a mound of coffee-grounds and there, by the beard of Owsley, stood an absolutely *dynamite* chick. Red hair, crazy blue eyes, and a protoplasmic distribution that made me think of a brick latrine. At the mere sight of this girl, certain physiological reactions overcame embarrassment and mortal terror.

I sure wished I had a pair of pants.

"What's happening?" she inquired around another giggle. *My God*, I thought, *she's from Long Island!* I decided to trust her.

"Well, see baby, I was makin' this run for my platoon, little smoke to sweeten the jungle, right? And, ah . . . I've gone a wee bit awry."

"Heavy." She jiggled sympathetically, and moved closer.

"Well, yeah, particularly since my C.O. don't like gettin' burned. Liable to amputate my ears is where it's at."

She smiled, and my eyes glazed. "No sweat. I can set you up."

"Right."

"No, really. I'm General Fonebone's old lady—I've got connections. I could probably fix you right up . . . if you weren't in *too* much of a hurry." She was *not* staring me in the eyes, and I made a few hasty deductions about General Fonebone's virility.

"I'm Jim Balzac. 'Balz' to you."

"I'm Suzy."

Six hours later I was back in the jungle. I had a pair of pants, some four and a half bricks from the General's private stash, a compass, two Dylan albums, and (although I was not to know it for weeks), a heavy dose of clap. I felt great, and it was all thanks to General Fonebone. If Suzy had not found life in Vietnam so boring, she would never have gone rummaging and uncovered the General's Secret Stash, a fell collection of strange tabs and arcane caps. She had induced me to swallow the largest single tab in the bunch, an immense purple thing with a skull embossed on it above the lone word: "HEAVY," and it appeared in retrospect to have been a triple tab of STP cut with ibogaine, benzedrine, coke and just a touch of Bab-O.

It might just as easily have been Fonebone's Own—the sensation was totally new to me. But it was certainly interesting. I experienced considerable difficulty in finding my mustache—which, of course, was right under my nose.

I could navigate without difficulty, after a fashion. But I discovered that I could whip up a ball of hallucinatory color-swirls in my mind, fire it like a cannonball, and watch it burst into a spiderweb of multicolored sparkles, as though an invisible protective shield two feet away walled me off from reality. With care, I could effect changes in the nature of the pulsing balls before they were fired, producing a variety of spectacular fireworks.

The jungle reared drunkenly above me. My outfit was straight ahead. I forged on, while in my darker crannies gonococcal viruses met and fell in love by the thousands, all unknown.

A particularly vivid splash of color caught my wandering attention; I had absently concocted a hellish color-ball of surpassing incandescence and detonated it. Its brilliant pattern hung before me a moment, as the rush took hold.

And then it very suddenly vanished.

I very nearly fell on my face. When I had my bearings again, I sent out another "shell." It burst pyrotechnically.

And as suddenly vanished. It made a noise best reproduced by inhaling sharply through clenched teeth while saying the word "Fffffup!"; vanished down behind a small hill ahead, *sucked* downward in a microsecond—only a stoned man could have divined the direction.

Something on the other side of that hill was eating my hallucinations.

I moved to the left like a stately zeppelin, caroming gently from the occasional tree. But I had two anchors dragging the ground, and before I got fifteen feet a tangled root brought me down with a crash.

And just before I hit, I saw something coming over the rise, and I knew that my mind had truly blown at last.

Coming toward me was a sixteen-foot-tall poached egg with pimples.

And then the lights—all those lights—went out.

Yteic-os moved from concealment, throbbing with astonished elation. No subtle attack was necessary, no cunning stimulus needed to elicit secretions of The

Force from this being. Heedless of danger, it radiated freely in all directions, idly expectorating energy-clusters as it walked.

Then Yteic-Os gasped (almost); for as it became aware of himher, it assumed a prone position, and disappeared. That is, its physical envelope remained, but all emanations ceased utterly; sentience vanished..

The Voracious One had no means of apprehending a subconscious mind. Such perverse deformities are extremely rare in the universe; heshe had in several billions of eons never chanced to so much as hear of such a thing. This led himher into a natural error: heshe assumed that these odd creatures emanated so incautiously because they had the ability to shut their minds off at will to escape absorption.

For, you see, thought is electrical in nature, and creative thought is akin to a short circuit, occurring when two unconnected thoughts arc together to form a totally new pattern. And such was Yteic-Os's diet.

And so heshe made a serious mistake. Heshe stealthily entered the empty caverns of Private Balzac's mind to try and restimulate life. Meanwhile, Yteic-Os's own nature and essence were laid open to the soldier's subconscious. One of the few compensations humans have for being saddled with such a clumsy nuisance as a subconscious mind is that these distorted clumps of semiawareness possess a passionate interest in survival. Balzac's subconscious remained hidden, probing, comprehending the nature of this novel threat. A nebulous plan of defense formed, was stored for the proper time. Yteic-Os searched in vain for Thought, while Thought watched him from ambush, and giggled.

Consciousness returned to Private Balzac with a jar and a "WHAAAAAT!?!" Yteic-Os, caught by surprise, flipped completely over on his back and rippled indignantly. This upstart would soon be only a belch—or something like one. The Voracious One licked hisher . . . well, you know what I mean.

* * *

"Whaaaaaat!?!"

I was awake. Somehow it had all been sorted out in my sleep: I didn't exactly know what the poached egg was, but I knew what it wanted to do. I thought I knew what to do about it. I would absolutely refuse to hallucinate, and starve it to death.

But I hadn't reckoned with the Terrible Tab I'd swallowed. I simply could *not* stop hallucinating! Colored whirlwinds and coruscating rainbows danced all around me like a mosaic in a Mixmaster; my eyeballs were prisms. Slowly the creative force of my mind was leaking away, being sucked into the egg before it could feed-back and regenerate itself.

I was being drained of originality, of wit, of inventiveness, of all the things that make life groovy. I had a grim vision of myself a few years hence, a short-haired square working in a factory living contentedly in Scarsdale with a frigid wife and a neurotic Pekingese, stumbling over the Cryptoquote in the *Daily News* and drinking Black Label before the TV. A grimmer vision I couldn't imagine, but I still missed it when, with a sucking sound, it disappeared into the poached egg.

It was quickly supplanted by other visions, however—but from the past rather than the future. To my utter horror, I realized that it was actually happening: my whole life was passing before my eyes, in little vignettes which were *slurped* up by the creature as fast as they formed.

In spite of myself I began watching them. In rapid succession I reviewed a lifetime of disasters: losing my transmission at the head of the Victory Parade, getting bounced out of bed a hair before climax when I accidentally called Betty Sue the wrong name, being violently ill on two innocent customers of Howard Johnson's . . .

Wait! A light-bulb rather unoriginally appeared over my head (and was eaten by the poached egg).

Howard Johnson's!!! My untimely nausea had come on my third day as a HoJo counterman, a direct result of the genius of Mr. Johnson himself. Early in his career, Johnson had hit upon the notion of urging all new employees to eat all the ice cream they wished, for free. He reasoned that they would soon become sick of ice cream, and hence cut employee pilferage from his overhead. The scheme had worked well for him and his heirs and assigns—why not for me?

Desperately I rammed my forebrain into low gear and cut in the afterburner. I dug into the tangled whorls of my cerebrum for all the creativity that heredity and environment had given me, and began to hallucinate as fast and as intricately as I could. I prayed that the poached egg would O.D.

Yteic-Os was caught in a quandary. The Force was radiating from this rococo little entity at an intolerable rate, and the creature would not stop projecting!

The Voracious One screamed—after hisher fashion—and tried frantically to assimilate the superabundance of food, to no avail. Even as heshe thrashed, desperately seeking to control hisher growth, heshe swelled, grew, expanded more and more rapidly, like a balloon inextricably linked to an air compressor. Heshe lost hisher egg shape, became round rather than ovoid, swelled, bloated to impossible dimensions, and—

—the inevitable happened.

And when I could see again, there was scrambled eggs all over the place.

I didn't hang around. Corporal Zeke was delighted to see me—it's embarrassing to have men under your command bumming joints from the enemy. But he was a little disappointed to learn that I only had four and a half bricks.

"That's okay, Corp," I assured him. "You guys can have my share. I'm straight for life."

"*What?*" gasped Zeke, shocked enough to deliver the first and only one-word speech of his life.

"Yep. After what I went through on the way over here, I'll never get stoned again as long as I live. Poached eggs eating hallucinations, cosmic invasion, Howard Johnson—it was just too intense, man, just too intense. A man who could freak out like that didn't ought to do dope. I've had a few bummers before, but I know when I've been warned."

Zeke was stupefied, but not so stupefied as to fail to try and change my mind. In subsequent weeks he went so far as to leave joints on my pillow, and once I caught him slipping hash into my K-rations. But like I say, I know when I've been warned, and you can't say I'm stupid.

I live a perfectly content life now that the war is over. Got me a wife, a nice little one-family in Scarsdale that I'll have entirely paid off in another twenty-five years, and a steady job down at the distributing plant—I get to bring home unlimited quantities of Black Label.

But sometimes I drink a little too much of it, and my wife Mabel says when I'm drunk—aside from becoming "disgustingly physical"—I often babble a lot. Something about having saved the world. . . .

I don't want to think about it.

TIDBIT: two more songs

(On this first song I wrote both the words and the lyrics. That is to say, I wrote it especially for this collection—the first song I've written in three years, if you don't count a birthday song I wrote for Jeanne last year and never sing publicly—and as yet I have been unable to come up with a tune. Feel free to make up your own.)

PERSPECTIVE

A cop with any decency at all looks like a hero
A millionaire knows billionaires who think that he's a zero
The shoes a lord rejected are a godsend to the churl
And an immie in the sewer looketh mighty like a pearl

A million people kill themselves attempting to be stars
While stars go nuts with loneliness and smoke the highest
  tars
Businessmen competing, and the ones that do the best
Win the hatred of their neighbor and a cardiac arrest

> So remember on those days when in your bed you
>   shoulda stood
> That somewhere there is someone who makes even
>   you look good
> It's only your perspective that has got you in a muddle
> You ain't too small a frog—you just been in too big a
>   puddle

When you're sittin' agonizin' over all the stuff you want
Just think about the *lucky* guy that got that stuff in front
The guy who's got it so together nothin' ever throws him
The guy that's so damn beautiful that no one ever knows
   him

His wealth is so enormous that he's all the time afraid
So heavy is he hung that he is hardly ever laid
So famous is his work that he will never top his last one
Everyone around him always tryin' to pull a fast one
(*chorus*)

That glass is not half-empty, man—to me it looks half-full
If it wasn't for the bullshit what would happen to the bull?
If you can tell me honestly your troubles is the worst
I'll take this song and eat it, friend, and you will be the
   first

I sorrowed for the shoeless man until I lost my feet
And grumbled at the rain until it started in to sleet
If you got eyes to see my song, and you got ears to hear it
Then you ain't at the bottom, baby: you ain't even near it
(*chorus*)

Mountain Lady

(Shortly after I met my future wife, Jeanne, on the North Mountain of Nova Scotia's Annapolis Valley, I was forced to leave the province and return to New York to try and settle an Immigration hassle. It took longer than I had hoped . . . and I found myself writing this song. When I got back, the first-draft lyrics sheet was pinned to the wall beside Jeanne's bed. Nowadays our household consists of me, Jeanne, and our daughter Luanna Mountainborne—for a total of one.)

MOUNTAIN LADY

(words and music by Spider Robinson)

Mountain Lady sing for me   Your singing makes me
glad to be alive
Mountain Lady bring to me your loving for it helps me to
survive
Mountain Lady stay with me, and let me drink your beauty
with my eyes
I want you to lay with me, and be there in the morning
when I rise
(chorus)

> You give me what I need, and you need what I can
> give
> Like you I live for loving and like me you love to live
> I swear I'll make you happy if there's any way I can
> And if you will be my Mountain Lady, I will be your
> man

Mountain Lady smile for me   Your smile is like the
rising of the sun
Wait a little while for me: I'm coming back as fast as I
can run
Mountain Lady talk with me, for talking is essential to our
growth
I want you to walk with me through all the days remaining
to us both

> You give me what I need, and you need what I can
> give
> Like you I live for loving and like me you love to live
> I swear I'll make you happy if there's any way I can
> And if you will be my Mountain Lady, I will be your
> man

Mountain Lady dance for me   Your dancing takes my
  breath away you know
Save a loving glance for me: I love it when you let your
  loving show
Mountain Lady give to me a kind of love I've never had
  before
I want you to live with me: I cannot live without you
  anymore

> You give me what I need, and you need what I can
>   give
> Like you I live for loving and like me you love to live
> My love is deep and stronger than a river running wild
> I want to be your lover, and the father of your child

MOUNTAIN LADY

10
TIN EAR

Call them Stargates if you want to. The term was firmly engraved in the public's mind, by science fiction writers with a weakness for grandiose jargon, fully fifty years before the first Spatial Anomaly was discovered and the War started. If you do call them Stargates, you probably call us Stargate Keepers, or Keepers for short.

But we call 'em 'Holes, for short, and we call ourselves Wipers.

It's all in how you look at it, of course. If we ever got to enter one, instead of just watching them and mopping up what comes out, we might have a different name for them—or if not, at least a different name for ourselves. ". . . and cheap ones, too," as the joke goes.

But the Enemy's drones keep popping out at irregular intervals, robot-destroyer planetoids with simple but straightforward programs written somewhere on the far side of hyperspace. So, in addition to the heroes who get to go after the source—and keep failing to return—somebody has to mount guard over every known 'Hole, to sound the alarm when a drone comes through, and hopefully to neutralize it (before it neutralizes *us*). The War is still, after twenty years, at the stage where intact prizes are more valuable than confirmed kills. Data outworth debris, and will for decades to come.

For the Enemy, apparently, as much as for us, or I

wouldn't be here. The first Enemy drone I ever saw could certainly have killed us both, if it had wanted to.

It was well that Walter and I inhabited separate Pods. We didn't get along at all. The only things we had in common were *(a)* an abiding hatred for the government which had drafted us into this sillyass suicidal employ (". . . before we had a chance to volunteer like gentlemen," we always added) and *(b)* a deep enjoyment of music.

But all Wipers share these two things. One of the few compensations our cramped and claustrophobic Pods feature are their microtape libraries and excellent playback systems (you can't read properly on combat status). And so it was possible for Walter and I to spend endless hours within the same general volume of space, listening to separate masterpieces over our headphones and arguing only occasionally. Walter had no sense of humor whatsoever, despised anyone who did, loathed any music of satirical, parodying or punning nature, and therefore was impossible to discuss music with. Or anything much.

But you can listen to a lot of good music if you have nothing else to do.

I was seventeen hours into Wagner's *Ring Des Niebelungen,* thoroughly exhausted but with the end in sight, when Walter's commlaser overrode my headphones. "George."

"*Wha?*" I yelled, but there was too much cacophony. We both had to kill our tapes. Damned if he didn't have *Siegfried* on himself, which annoyed me—I was certain, without asking, that he liked Wagner for all the wrong reasons.

"Alert status," he said, yanking me from music back to reality.

"Right." I slapped switches and reached out to touch my imitation rabbit's foot. So the 'Hole was puckering up, eh? A noble death might lie seconds away. With all possible speed I joined Walter in training all the considerable firepower we possessed on the 'Hole.

And the bastard popped out a couple thousand miles *to one side of* the 'Hole and bagged us both. Unheard of; still unexplained. Even Abacus Al, the computer you can count on, was caught flat-footed. Tractor beam grabs me, *clang!*, reels in fast, *CLANG!*, half a billion Rockies' worth of Terran hardware on alien flypaper, *slump*, body goes limp in shock-webbing, *ping!*, lights go out.

"George," Walter was saying in my headphones, "are you all right?"

"I'll see," I replied, but by then some sort of anti-laser device must've been interposed by the drone-planetoid which held us captive, for the headphones went dead. I sighed and checked my Pod. It was on its gyrostabilized tail, "upright." All my video screens were dead, except for the one that showed me about twenty degrees of starry space straight "overhead"—my location with reference to Walter was unknown. This was serious if I intended to live, which I did. But before I tried the radio I inspected my weapons control systems (dead in all directions except "up"), main drive (alive, but insufficient to pry me loose), and my body (alive and apparently unharmed). *Then* I heated up the radio on standard emergency band.

"Down one freak, Cipher A," I said crisply and quickly, getting it all out before static jammed that frequency. Then I dialed 'er down to the next fre-quency on the "standard" list, instructed Abacus Al the AnaLogic to convert to Cipher A before transmitting. "Walter?"

"Here." Flat, mechanical voice—Al's rendition of human speech, just like what Walter was hearing from me.

"Simpleton machine."

"Yah."

"Capture, not kill. Programmed to immobilize us, disarm us, blind us, and prevent meaningful communication between us. As soon as it dopes out Cipher A, it'll . . ."

A million pounds of frying bacon drowned me out. I dropped freak by the same interval again and shifted to Cipher B, allegedly a much tougher cipher to break. They call it "the best nonperfect cipher possible."

Walter was waiting on the new freak. "It's essential," he began at once, "that we determine whether this drone-planetoid is a Mark I or a Mark II."

"Damn right," I agreed. "If we can work out our relative positions we've at least got options."

And a roar of static threw Cipher B out the window.

Both types of Enemy planetoid have only the two tractor beams—but the relative *positions* of them are one of the chief distinguishing features from the outside. If this was a Mark I, we could both throw full power to our drives—and while they wouldn't be sufficient to peel us loose, their energies should cross, like surgical paired-lasers, at the center of the planetoid, burning out its volitional hardware. If this were Mark the Second, the same maneuver would have our drives cross in the heart of the power-plant and distribute the component atoms of all three of us across an enormous spherical volume of space. But how could we compute our positions blind, on a sphere with no agreed-upon poles or meridians anyhow, and communicate them to each other's computers without tripping the damned planetoid's squelch-program? The cagey son of a bitch had cracked Cipher B too easily—apparently it was programmed to jam anything that it computed to be "exchange of meaningful information" whether it could decipher it or not. That suggested that Cipher C, the Perfect Cipher might be the only answer.

The perfect cipher (really a code-cipher) was devised way back in the 1900s, and has never been improved upon. You have a computer generate an *enor-*

mous run of random numbers, in duplicate. You give a
copy of the printout to each communicator, and down
the column of random numbers they go, each writing
out the alphabet, one letter to each number, over and
over again. For each successive letter they want to en-
cipher and send, they jump down to the next
alphabet-group in line, select the random number adjacent to the desired letter, and transmit that number.
A savvy AnaLogic deduces pauses, activates voder:
communication. The cipher *cannot* be broken by anyone not in possession of an identical list of random
numbers, for it produces utterly no pattern. (We had
a code, by the way, a true code, in which prearranged
four-letter groups stood for various prearranged
phrases. But not a phrase on the list applied to our
situation—I love the Army—and using a series of ex-
clusively four-letter groups would have tipped off the
alien computer that a code was in use.)

But Cipher C had one flaw that I could see, and so
I hesitated before dialing the frequency again. If we
lost *this* chance, we were effectively deaf and dumb
as well as blind. *Oh God,* I prayed, *give Walter just
this once, and for no more than fifteen minutes, at
least half a brain.* I dialed the new freak.

". . . got to take starsights," he was saying. "It's the
only way to . . ."

"SHUT UP!"

"Eh?"

"No sound. Listen. *Heed.* Okay? *Care*fully. Yes,
'sights,' but do not under any circumstances repeat any
phrase or word-group I use. *Comprende?*"

I breathed a silent prayer.

"Why shouldn't I repeat any phrase or word-group
you use?" Walter asked, puzzlement plain even
through voder.

"GODDAMMIT," I roared, but I was addressing
only another roar of static. Groups with identical
numbers of characters, in repeated sequence, were the

only clue the Enemy computer had needed. It was "meaningful communication," so it was jammed.

One more standard band left on the list. If we had to hunt for each other on offbeat frequencies, it could take forever to establish contact.

I scratched a telemetry contact and consulted Abacus Al. "How," I programmed, "can I communicate meaningful information without communicating meaningful information?"

That's the kind of question that makes most computers self-destruct, like an audio amplifier with no output connected. But Al is built to return whimsy with whimsy, and his sense of humor is as subtle as my own. "WRITE A POEM," he replied, "OR SING A SONG."

I snorted.

"No good," I punched. "Can't use words."

"HUM," Al printed.

A nova went off in my skull.

I crosswired the microtape library in Al's belly to the radio in his rump, and had him activate the last standard frequency. It was live but silent: Walter had finally figured out his previous stupidity. He waited for me to come up with inspiration this time.

I keyed the opening bars of an ancient Beatles' song. "We Can Work It Out." In clear. And then killed it before the melody repeated.

A long silence, while Walter slowly worked it out in his thick head. *Come on, dummy*, I yelled in my head, *give me something to work with!*

And my headphones filled with the strains of the most poignant song from *Cabaret*: "Maybe This Time."

Thank God!

I keyed Al's starchart displays and thought hard. The chunk of sky *I* saw was useless unless I could learn what Walter was seeing over his *own* head—the two combined would give us a fix. I couldn't see the 'Hole, and I had to assume he couldn't either, or he'd have surely mentioned it already.

Or would he? Anyone with half a brain would have . . .

I keyed in the early twenty-first-century Revivalist dirge, "Is There a Hole in Your Bucket?" and hoped he wouldn't think I was requesting a damage report.

He responded with the late twenty-first-century anti-Revivalist ballad, "The Sky Ain't Holy No More."

Okay, then. Back to the Beatles. "Tell Me What You See."

Walter paused a long time, and at last gave up and sent the intro to Donald MacLean's Van Gogh song—the line that goes, "Starry, starry night . . ." He was plainly stymied.

Hmmm. I'd have to think for both of us.

Inspiration came. I punched for a late twenty-first-century drugging-song called—"Brother Have You Got Any Reds?" There were few prominent red stars in this galactic neighborhood—if any appeared in Walter's "window" it might help Al figure our positions.

His uptake was improving; the answer was immediate. Ellington's immortal: "I Ain't Got Nothin' But the Blues."

So much for that one.

I was stumped. I could think of no more leading questions to ask Walter with music. If he couldn't, for once, make his own mind start working in punny ways, we were both sunk. Any time now, real live Enemies might pop out of the 'Hole, and there was no way of telling what they were like, because no human had ever survived a meeting with them at that time. *Come on, Walter.*

And he floored me. The piece he selected almost eluded me, so obscure was it: an incredibly ancient children's jingle called, "The Bear Went Over the Mountain."

I studied the starcharts feverishly, trying to visualize the geometry ("cosmometry?")—I lacked enough skill to have Al do it for me. If Walter could see the Bear at all, it seemed to me . . ."

I sent the chorus of "Smack Dab in the Middle," the legendary Charles's version, and hoped Walter could sense the question mark.

Again, his answer baffled me momentarily—another Beatles song. *He loves me?* I thought wildly, and then I got it. "Yeah yeah yeah!"

My fingers tickled Abacus Al's keys, a ruby light blinked agreement, and Al's tactical assessment appeared on the display.

MARK ONE, it read.

"Walter," I yelled in clear, "Main drive. *Now!*"

And so when the *live* Enemies came through the 'Hole, *we* had the drop on *them*, which is how man got his first alien corpses to study, which is why we're (according to the government) winning the War these days. But the part of the whole episode that I remember best is when we were waiting there dead in space—in ambush—our remaining weaponry aimed at the 'Hole, and Walter was saying dazedly, "The most amazing thing is that the damned thing just sat there listening to us plot its destruction, with no more sense of self-preservation than the foresight of its programmers allowed. It just sat there . . ."

He giggled—at least, from anyone else I'd have called that sound a giggle.

". . . sat there the . . . the whole time . . ."

He was definitely giggling now, and it must be racial instinct because he was doing it right.

". . . the . . . the whole time just . . ."

He lost control and began laughing out loud.

"Just *taking notes*," he whooped, and I dissolved into shuddering laughter myself. Our mutual need for catharsis transformed his modest stinker into the grandest pun ever made, and we roared. Even Abacus Al blinked a few times.

"Walter," I said, "I've got a feeling the rest of this hitch is going to be okay."

And then alarms were going off and we went smoothly into action as a *unit*, and the Enemy never had a chance.

TIDBIT: foreword to "The Magnificent Conspiracy"

I consider myself a member of a magnificent conspiracy, and I am attempting to recruit anyone I can— you, if you're not busy. There's a whole lot of us, more than you might think, and our stated purpose is to save the world.

This requires a conspiracy to smuggle knowledge, to disseminate some simple truths that no one taught us in school. Basic keys to how the Universe works, which are not so much suppressed as buried in misinformation and derision.

For instance: If there's one thing I absorbed through the skin from better than ninety percent of the teachers I ever had, virtually all adults who spoke with me, and all the entertainment media I was ever exposed to, if there's one thing my upbringing prepared me to accept as *certain*, it is the proposition that work of any kind is a drag. That the smart man avoids work, that the dummies are the ones who *work* for a living, that leisure is the proper pursuit of the clever and the powerful.

Isn't that incredible? It took me better than a quarter century to learn, the hard way, that hard work at something you want to be doing is the most fun that you can have out of bed (and that working at something you don't want to be doing is a logical impossibility—that we are all self-employed). To learn that the dummies are the ones who think it possible to cheat the boss or the customers without cheating themselves; to learn that the smart man finds ways to make everything he does be work; to learn that "leisure" time is truly pleasurable (indeed tolerable) only to the extent that it is subconscious grazing for information with which to infuse newer, better work.

They told me often, for instance, that "marriage is hard work"—but somehow the way they said it made that sound like a disadvantage.

How could I have been so basically misinformed for so many years about the way reality is put together? Why did I have to *deduce,* from three decades worth of memories (breaking my back helping my neighbor David get his hay in before the rain, dodging pitchforks in pitch darkness on top of a truckload of hay; literally writing myself into unconsciousness to meet the deadline for the *Stardance* novel; shoveling out my outhouse in the summertime), that the definition of "riches" is "abundant meaningful work"?

Why didn't anybody *tell* me?

Why did I go through years of thinking I was weird before I found out that *everyone* is an empath, that *all* humans are at least potentially telepathic, only some are afraid or unable to admit it? How did they manage to make "you reap what you sow" such a trite-sounding cliche that I'm only just now realizing that I can *never* get away with *anything,* and never have? For what twisted reason were I and all my generation told so often of the essential corruption and evil of Man that it took me twenty rs to learn to like myself and another ten to begin to love myself?

The protagonist of this next and final story asks a direct question in the story's last sentence. It does not, obviously, receive an answer.

It may, someday, in the novel-length expansion of this story which I want to write. At present I cannot seem to *sell* it, though—editors tell me it "ain't commercial—it's just too way out." Doomful apocalyptic visions are apparently more plausible than optimistic projections these days. God help us.

In the meantime, though, I am quite content to leave the final question unanswered—because I want *you* to answer it. I want you to indulge in the most fundamental kind of wishful thinking: imagine yourself in the protagonist's place, with the opportunity he has just been offered.

What would *you* do with it?

If I can get you to fantasizing along those lines, the next thing you know you'll be in the conspiracy, too.

11
THE MAGNIFICENT CONSPIRACY

<hr>

1.

By the time I had pulled in and put her in park, alarm bells were going off all over my subconscious, so I just stayed put and looked around. After a minute and a half, I gave up. *Everything* about the place was wrong.

Even the staff. Reserved used-car salesmen are about as common as affable hangmen—but I had the whole minute and a half to myself, and as much longer as I wanted. The man semivisible through the dusty office window was clearly aware of my arrival, but he failed to get up from his chair. So I shut off the ignition and climbed out into unair-conditioned July, and by God even the music was wrong. It wasn't Muzak at all; it was an old Peter, Paul & Mary album. How can you psych someone into buying a clunker with music like that?

Even when I began wandering around kicking tires and glancing under hoods he stayed in the office. He seemed to be reading. I was determined to get a reaction now, so I picked out the classiest car I could see (easily worth three times as much as my Dodge), hot-wired her and started her up. As I'd expected, it fetched him—but he didn't hurry. Except for that, he was standard-issue salesman—which is like saying, "Except for the sun-porch, it was a standard issue fighter jet."

"Sorry, mister. That one ain't for sale."

I looked disappointed. "Already spoken for, huh?"

"Nope. But you don't want her."

I listened to the smooth, steady rumble of the engine. "Oh, yeah? Why not? She sounds beautiful."

He nodded. "Runs beautiful, too—now. Feller sold it to us gimmicked 'er with them pellets you get from the Whitney catalog. Inside o' five hundred miles you wouldn't have no more rings than a spinster."

I let my jaw drop.

"She wouldn't even be sittin' out here, except the garage is full up. Could show you a pretty good Chev, you got your heart set on a convertible."

"Hey, listen," I broke in. "Do you realize you could've kept your mouth shut and sold me this car for two thousand flat?"

He wiped his forehead with a red handkerchief. "Yep. Couple year ago, I would've." He hitched his glasses higher on his nose and grinned suddenly. "Couple year ago I had an ulcer."

I had the same disquieting sensation you get in an earthquake when the ground refuses to behave properly. I shut the engine off. "There isn't a single sign about the wonderful bargains you've got," I complained. "The word 'honest' does not appear anywhere on your lot. You don't hurry. I've been here for three minutes and you haven't shaken my hand and you haven't tried to sell me a thing and *you don't hurry*. What the hell kind of used-car lot *is* this?"

He looked like he was trying hard to explain, but he only said "Couple of year ago I had an ulcer," again, which explained nothing. I gave up and got out of the convertible. As I did so, I noticed for the first time an index card on the dashboard which read "$100."

"That can't be the price," I said flatly. "Without an *engine* she's worth more than that."

"Oh, no," he said, looking scandalized. "That ain't the price. Couldn't be: price ain't fixed."

Oh. "What determines the price?"

"The customer. What he needs, how bad he needs it, how much he's got."

This of course is classic sales-doctrine—but you're

not supposed to *tell* the customer. You're supposed to go through the quaint charade of an asking price, then knock off a hastily computed amount because "I can see you're in a jam and I like your face."

"Well then," I said, trying to get this script back on the track, "maybe I'd better tell you about my situation."

"Sure," he agreed. "Come on in the office. More comfortable there. Got the air conditioning."

I saw him notice my purple sneakers as I got out of the convertible—which pleased me. You can't buy them that garish—you have to dye them yourself.

And halfway to the office, my subconscious identified the specific tape being played over the sound system.

Just a hair too late; the song hit me before I was braced for it. I barely had time to put my legs on automatic pilot. Fortunately, the salesman was walking ahead of me, and could not see my face.

Album 1700, side one, track six: "The Great Mandella (The Wheel of Life)."

So I told him
That he'd better
Shut his mouth
And do his job like a man
And he answered
Listen (*father didn't even come to the funeral and the face in the coffin was my own but oh God so thin and drawn like collapsed around the skull and the skin like gray paper and the eyes dear Jesus the eyes he looked so content so hideously* content *didn't he understand that he'd blown it blown it*) own it very long, Mr. Uh?"

He was standing, no, squatting by my Dodge, peering up the tailpipe. The hood was up.

If you're good enough, you can put face and mouth on automatic pilot, too. I told him I was Bob Campbell, and that I had owned the Dodge for three years. I told him I was a clerk in a supermarket. I told him I had a wife and two children and an M.A. in Business Administration. I told him that I needed a newer-model car to try for a better job. It was a plausible story; he didn't seem to find anything odd about my facial expressions, and I'm sure he believed every word. By the time I had finished sketching my income and outgo, we were in the office and the door was closing on the song:

Take your place on
The Great Mandala
As it moves through your brief moment of (click) time that Dodge of yours had a ring job, too, Bob."

I came fully aware again, remembered my purpose.

"Ring job? Look, uh . . ." we seated ourselves.

"Arden Larsen."

"Look, Arden, that car had a complete engine overhaul not five thousand miles ago. It's . . ."

"Stow it, Bob. From the inside of your exhaust pipe alone my best professional estimate is that you are getting about forty or fifty miles to a quart of oil. Nobody can overhaul a slant-six that bad." I began to protest. "If that engine was even so much as steam-cleaned less'n ten thousand mile ago I'll eat my socks."

"Just a damned minute, Larsen . . ."

"Don't ever try to bamboozle a used-car man my age, son—it just humiliates the both of us. Now, it's hard to tell for sure without jackin' up the front end or drivin' her, but I'd guess the actual value of that Dodge to be about a hundred dollars. That's half of what it'd cost you to rent a car for as long as the Dodge is liable to last."

"Well, of all the colossal . . . ! I don't have to listen to this crap!" I got up and headed for the door, which was a bit corny and a serious mistake, because when I was halfway to the door he hadn't said a word and when I was upon it he still hadn't said a word and I was so puzzled at how I could have overplayed it so badly that I actually had the door open before I remembered what lay outside it:

Tell the jailer
Not to bother
With his meal of bread and water today

He is fasting till the killing's over here and I'll get you some ice water, Bob. Must be ninety-five in the shade out there. You'll be okay in a minute."

"Yeah. Sure." I stumbled back to my seat and gratefully accepted the ice water he brought from the refrigerator in a corner of the office. I remembered to keep my back very straight. *Get a hold on yourself, boy. It's just a song. Just some noise . . .*

"Now as I was sayin', Bob . . . figure your car's worth a hundred. Okay. So figure the Dutchman up the road'd offer you two hundred, and then sell it to some sorry son of a bitch for four. Okay. Figure if you twisted his arm, he'd go three—Mid-City Motors in town'd go that high, just to get you offa the lot quick. Okay. So I'll give you four and a quarter."

I sprayed icewater and nearly choked. "Huh?"

"And I'll throw in that fancy convertible for three hundred, if you really want her—but you'll have to let us do the ring job first. Won't cost you anything, and I could let you have a loaner 'til we get to it. Oh yeah, an' that $100 tag you was askin' about is our best estimate of monthly gas, oil and maintenance outlay. I'd recommend a different car for a man in your situation myself, but it's up to you."

I didn't have to pretend surprise; I was flabbergasted. "Are you out of your *mind*?" Apparently my employer was given to understatement.

He didn't have the right set of wrinkles for a smile like that; he must have just learned how. "Feels like I get saner every day."

"But . . . but you can't be serious. This is a rib, right?"

Still smiling, he pulled out a wallet the size of a paperback dictionary and counted out one hundred and twenty-five dollars in twenties and fives. He held it out in a hand so gnarled it looked like weathered maple. "What do you say? Deal?"

"I say, 'You're getting reindeer-shit all over my roof, fatso.' What's the catch?"

"No catch."

"Oh, no. You're offering me a free lunch, and I'm supposed to just fasten the bib and open my mouth, right? Is that convertible hot, or what?"

He sighed, scratched behind his glasses. "Bob, your attitude makes sense, in a world like this. That's why I don't much like a world like this, and that's why I'm working here. Now I understand how you feel. I've seen ten dozen variations of the same reaction since I started working for Mr. Cardwell, and it makes me a little sadder every time. That convertible ain't hot, and there ain't no other catch neither. I'm offerin' you the car for what she's honestly worth, and if you can't believe that, why, you just go down the line and see the Dutchman. He'll skin you alive, but he won't upset you any."

I know when people are angry at me. He was angry, but not at me. So I probed.

"Larsen, you've got to be completely crazy."

He blew up.

"You're damn right I am! Crazy means out o' step with the world, and accordin' to the rules o' the world I'm supposed to cheat you out of every dime I smell on ya plus ten percent an' if you like that world so much that you wanna subsidize it then you get yer ass outa here an' go see the Dutchman but what*ever* you do don't you tell him we sent ya you got that?"

Nothing in the world makes a voice as harsh as the shortness of breath caused by a run-on sentence. I waited until he had fed his starving lungs and then said, "I want to see the manager," and he emptied them again very slowly and evenly, so that when he closed his eyes I knew he was close to hyperventilating. He clenched his fingers on the desk between us as though he were trying to pull it toward him, and when he opened his eyes the anger was gone from them.

"Okay, Bob. Maybe Mr. Cardwell can explain it to you. I ain't got the right words."

I nodded and got up.

"Bob . . ." He was embarrassed now. "I didn't have no call to bark at you thataway. I can't blame you for bein' suspicious. Sometimes I miss my ulcers myself. It's—well, it's a lot easier to live in a world of mud if you tell yourself there ain't no such thing as dry land."

It was the first sensible thing he'd said.

"What I mean, I'm sorry."

"Thanks for the ice water," I said.

He relaxed and smiled again. "Mr. Cardwell's in the garage out back. You take it easy in that heat."

I knew that I'd stalled long enough for the cassette or record or whatever it was to have ended, but I treated the doorknob like an angry rattlesnake just the same. But when I opened it, the only thing that hit me in the face was the hot dry air I'd expected. I left.

2.

I went through an arched gate in the plank fence that abutted the office's rear wall, and followed a wide strip of blacktop through weedy flats to the garage.

It was a four-bay job, a big windowless wood building surrounded with the usual clutter of hand-trucks, engine blocks, transmissions, gas cans, fenders, drive trains, and rusted oil drums. All four bays were closed, in spite of the heat. It was set back about five hundred yards from the office, and the field behind it was lushly overgrown with dead cars, a classic White Elephant's Graveyard that seemed better tended than most. As I got closer I realized the field was actually organized: a section for GM products, one for Chryslers, one for Fords and so on, each marked with a sign and subdivided by model and, apparently, year. A huge Massey-Ferguson sat by one of three access roads, ready to haul the next clunker in to its appointed resting place. There was big money in this operation, very impressive money, and I just couldn't square that with Arden Larsen's crackpot pricing policy.

Arden seemed to have flipped the cassette to side two of *Album 1700*. I passed beneath a speaker that said it dug rock and roll music, and entered the garage through a door to the right of the four closed bays. Inside, I stopped short. Whoever heard of an air-conditioned garage? Especially one this size.

Big money.

Over on the far side of the room, just in front of a Rambler, the floor grew a man, like the Wicked Witch melting in reverse. It startled the hell out of me—until I realized he had only climbed out of one of those rectangular pits the better garages have for jobs where a lift might get in the way. With the help of unusually efficient lighting, I studied him as he approached me.

Late fifties, snow white hair and goatee, strong jaw and incongruously soft mouth. A big man, reminding me strongly of Burl Ives, but less bulky, whipcord fit. An impression of enormous energy, but used only by volition—he walked slowly, clearly because he saw no need to hurry. Paradoxical hands: thin-fingered and aristocratic, but with the ground-in grime which is the unmistakable trademark of the professional or dedicated-amateur mechanic. The right one held a pipe-wrench. His overalls were oily and torn, but he wore them like a not-rented tux.

I absorbed and stored all these details automatically, however, while most of my attention was taken up by the utter *peacefulness* of his face, of his eyes, of his expression and carriage and manner. I had never seen a man so manifestly content with his lot. It showed in the purely decorative way in which the wrinkles of his years lay upon his face; it showed in the easy swing of his big shoulders and the purposeful but carefree stride; it showed in the eager yet unhurried way that his eyes measured me: not as a cat sizes up another cat, but as a happy baby investigates a new person—with delighted interest. My purple sneakers *pleased* him. He was plainly a man who drank of his life with an unquenchable thirst, and it annoyed the hell out of me, because I knew good and goddam well when was the last time I had seen a man possessed of such peace and because nothing on earth was going to make me consciously acknowledge it.

But I am not a man whose emotions are wired into his control circuits. I smiled as he neared, and my body language said I was confused, but amiably so.

"Mr. Cardwell?"

"That's right. What can I do for you?" The way he asked it, it was not a conversational convention.

"My name's Bob Campbell. I . . . uh . . ."

His eyes twinkled. "Of course. You want to know if Arden's crazy, or me, or the both of us." His lips smiled, then got pried apart by his teeth into a full-blown grin.

"Well . . . something like that. He offered to buy my car for uh, more than it's worth, and then he offered to sell me the classiest-looking car on the lot for . . ."

"Mr. Campbell, I'll stand behind whatever prices Arden made you."

"But you don't know what they are yet."

"I don't need to," he said, still grinning. "I know Arden."

"But he offered to do a free ring job on the car, for Chrissake."

"Oh, that convertible. Mr. Campbell, he didn't do that 'for Chrissake'—Arden's not a church-going man. He did it for his sake, and for mine and for yours. That car isn't worth a thing without that ring job—the aggravation it'd give you would use up more energy than walking."

"But—but," I sputtered, "how can you possibly survive doing that kind of business?"

His grin disappeared. "How long can any of us survive, Mr. Campbell, doing business any other way? I sell cars for what I believe them to be genuinely worth, and I pay much more than that for them so that people will sell them to *me*. What's wrong with that?"

"But how can you make a profit?"

"I can't."

I was shocked speechless. When he saw this, Cardwell smiled again—but this time it was a smile underlaid with sadness.

"Money, young man, is a symbol representing the life energy of those who subscribe to it. It is a useful and even necessary symbol—but because it is only a symbol, it is possible to amass on paper more profit than there actually is to be made. The more people who insist on making a profit, all the time, in every dealing, the more people who will be required to go bankrupt—to pour their life-energy into the system and get nothing back—in order to keep the machine running. A profit is without honor, save in its own country—there is certainly nothing sacred about one. Especially if you don't need it."

I continued to gape.

"Perhaps I should explain," he went on, "that I was born with a golden spoon in my mouth. My family has been unspeakably wealthy for twelve generations, controlling one of the oldest and most respected fortunes in existence—the kind that calls for battalions of tax lawyers in every country in the world. My personal worth is so absurdly enormous that if I were to set a hundred dollar bill on fire every minute of my waking life I would never succeed in getting out of the highest income tax bracket."

"You . . ." My system flooded with adrenaline. "You *can't* be *that* Cardwell."

BIG money.

"There are times when I almost wish I wasn't. But since I have no choice at all in the matter, I'm trying to make the best of it."

"By throwing money away?" I yelped, and fought for control.

"No. By putting it back where it belongs. I inherited control of a stupendous age-old leech—and I'm forcing it to regurgitate."

"I don't understand." I shook my head vigorously and rubbed a temple with my thumb. "I just don't understand at all."

He smiled the sad smile again, and the pipe-wrench loosened in his grip for the first time. "You don't have to, you know. You can take your money from Arden and drive home in a loaner and pick up your convertible in a few days and then put it out of your mind. All I'm selling is used cars."

He was asking me a question.

I shook my head again, more slowly. "No . . . no, I'd like to understand, I think. Will you explain?"

He put the wrench down on an oil drum. "Let's sit down."

There were a pair of splendidly comfortable chairs in the rear of the garage, with foldaway armrests that let you select for comfort or elbow room at need. Beyond them stood an expensive (but not frost-free) refrigerator, from which Cardwell produced two frosty cold bottles of Dos Equis. I accepted one and sat in the nearer chair, ignoring the seductive comfort of its reclining back and keeping my spine as straight as I had in the office. Cardwell sprawled back in his and put his feet up on a beheaded slant-six, and when he drank from his Dos Equis he gave it his full attention.

I regret to say I didn't. *Despite* all the evidence, I could not make myself believe that this grease-stained mechanic with his sneakers on an engine block was actually THE Raymond Sinclair Cardwell. If it was true, my fee was going to quintuple, and Hakluyt was fucking well going to pay it. Send a man after a cat, and forget to mention that it's a black panther . . . *Jesus.*

Cardwell's chair actually had a beverage-holder built tastefully into the armrest; he set his beer in it and folded his arms easily. He spoke slowly, thoughtfully; and he had that knack of observing you as he spoke, modifying his word-choice by feedback. I have the knack myself; but I wondered why a man in his situation would have troubled to acquire it.

I found myself trying as hard to understand him as he was trying to be understood.

I don't know (he said) if I can convey what it's like to be born preposterously wealthy, Mr. Campbell, so I won't try. It presents one with an incredible view of reality that cannot be imagined by a normal human being. The world of the very rich is only tangentially connected with the real world, for all that their destinies are intertwined. I lived totally in that other world and that world-view for thirty-six years, happily moving around mountains of money with a golden bulldozer, stoking the fires of progress. I rather feel I was a typical multibillionaire, if that conveys anything to you. My only eccentricity was a passion for working on cars, which I had absorbed in my youth from a chauffeur I admired. I had access to the finest assistance and education the world had to offer, and became rather handy. As good as I was with international finance and large-scale real estate development and interlocking cartels and all the other avenues through which a really enormous fortune is interconnected with the world, I enjoyed manipulating my fortune, *using* it—in some obscure way I believe I felt a duty to do so. And I *always* made a profit.

It was in London that it changed.

I had gone there to personally oversee a large and

complex merger involving seven nations. The limousine had just left the airport when the first shot killed my driver. He was the man who taught me how to align-bore an engine block and his name was Ted. The window was down; he just hurled sideways and soiled his pants. I think I figured it out as the second shot got my personal bodyguard, but by then we were under the wheels of the semi. I woke up eight weeks later, and one of the first things I learned is that no one is ever truly unconscious. I woke up speaking in a soft but pronounced British accent precisely like that of my private nurses, and it persisted for two days.

I discovered that Phillip, the bodyguard, had died. So had Lisa, a lady who meant entirely too little to me. So had Teal, the London regional director who had met my plane, and the driver of the semi. The rifleman had been apprehended: a common laborer, driven mad by his poverty. He had taken a gun to traffic in the same way that a consistently mistreated Doberman will attack anyone who approaches, because it seemed to him the only honorable and proper response to the world.

Cardwell drank deep from his beer.

My convalescence was long. The physical crisis was severe, but the spiritual trauma was infinitely greater. Like Saint Paul, I had been smashed from my horse, changed at once from a mover and shaper to a terrified man who hurt terribly in many places. The best drugs in the world cannot truly kill pain—they blunt its edge without removing it, or its terrible reminder of mortality. I had nearly died, and I suddenly had a tremendous need to explain to myself why that would have been such a tragedy. I could not but wonder who would have mourned for me, and how much, and I had a partial answer in the shallow extent of my own mourning for Ted and Phillip and Teal and Lisa. The world I had lived my life in was one in which there was little love, in which the glue of social relationships was not feelings, but common interests. I had narrowly, by the most costly of medical miracles, avoided inconveniencing many hundreds of people, and not a damn thing else.

And, of course, I could not deal with this consciously or otherwise. My world-view lacked the "spiritual vocabulary" with which to frame these concepts: I desperately needed to resolve a conflict I could not even express. It delayed my effective recovery for weeks beyond the time when I was technically "on my feet"—I was simply unable to reenter the lists of life, unable to see why living was worth the terrible danger of dying. And so my body healed slowly, by the same instinctive wisdom with which it had kept my forebrain in a coma until it could cope with the extent of my injuries.

And then I met John Smiley.

Cardwell paused for so long that I had begun to search for a prompting remark when he continued.

John was an institution at that hospital. He had been there longer than any of the staff or patients. He had not left the bed he was in for twelve years. Between his ribcage and his knees he was mostly plastic bags and tubes and things that are to a colostomy bag what a Rolls-Royce is to a dogcart. He needed one and sometimes two operations every year, and his refusal to die was an insult to medical science, and he was the happiest man I have ever met in my life.

My life had taught me all the nuances of pleasure; joy, however, was something I had only dimly sensed in occasional others and failed to really recognize. Being presented with a pure distillate of the thing forced me to learn what it was—and from there it was only a short step to realizing that I lacked it. You only begin to perceive where you itch when you learn how to scratch.

John Smiley received the best imaginable care, far better than he was entitled to. His only financial asset was an insurance company which grudgingly disbursed enough to keep him alive, but he got the kind of service and personal attention usually given only to a man of my wealth. This puzzled me greatly when I first got to know him, the more so when I learned that he could not explain it himself. But I soon understood.

Virtually every doctor, nurse, and long-term patient in the hospital worshipped him. The rare, sad few who would have blackly hated him were identified by the rest and kept from him. The more common ones who desperately needed to meet him were also identified, and sent *to* him, subtly or directly as indicated.

Mr. Campbell, John Smiley was simply a fountain of the human spirit, a healer of souls. Utterly wrecked in body, his whole life telescoped down to a bed he didn't rate and a TV he couldn't afford and the books scrounged for him by nurses and interns and the Pall Malls that appeared magically on his bedside table every morning—and the people who chanced to come through his door—John made of life a magnificent thing. He listened to the social and sexual and financial and emotional woes of anyone who came into his room, drawing their troubles out of them with his great gray eyes, and he sent them away lighter in their hearts, with a share of the immeasurable joy he had somehow found within himself. He had helped the charge nurse when her marriage failed, and he had helped the head custodian find the strength to raise his mongoloid son alone, and he had helped the director of the hospital to kick Demerol. And while I knew him, he helped a girl of eighteen die with grace and dignity. In that hospital, they sent the tough ones around, on one pretext or another, to see John Smiley—and that was simply all it took.

He had worked for the police as a plainclothesman, and one day as he and his partner were driving his own car into the police garage, a two-ton door had given way and come down on them. Ackroyd, his partner, had been killed outright, and so Mrs. Ackroyd received an award equivalent to half a million dollars. John's wife was less fortunate—his life was saved. They explained to her that under the law she would not collect a cent until he was dead. Then they added softly that they gave him a month at the outside. Twelve years later he was still chain-smoking Pall Malls and bantering with his wife's boyfriend when they came to visit him, which was frequently.

I wandered into John Smiley's room one day, sick in my heart and desperately thirsty for something more than thirty-six years had taught me of life, seeking for a reason to go on living. Like many others before and since, I drank from John Smiley, drank from his seemingly inexhaustible well of joy in living—and in the process, I acquired the taste.

I learned some things.

Mostly, I think, I learned the difference between pleasure and joy. I suppose I had already made the distinction, subconsciously, but I considered the latter a fraud, an illusion overlaid upon the former to lend it respectability. John Smiley proved me wrong. His pleasures were as restricted as mine had been unrestricted—and his joy was so incandescently superior to mine that on the night of the day I met him I found myself humming the last verse of "Richard Corey" in my mind.

Cardwell paused, and his voice softened.

He forgave me my ignorance.

He forgave me my money and my outlook and my arrogance and *treated me as an equal*, and most amazing of all, he made me forgive myself.

The word "forgive" is interesting. Someone robs you of your wallet, and they find him down the line and bring him back to you, saying, "We found your wallet on this man," and you say, "That's all right. He can have—can have had—it; I fore-give it to him."

To preserve his sanity, John Smiley had been forced to "fore-give" virtually everything God had given him. In his presence you could not do less yourself.

And so I even gave up mourning a "lost innocence" I had never had, and put the shame he inspired in me to positive use. I began designing my ethics.

I interrupted for the first and last time.

"A rich man who would design his own ethics is a dangerous thing," I said.

Damn right (he said, with the delight of one who sees that his friend really *understands*). A profit is without honor except in its own country—but that's a hell of a lot of territory. The economic system reacts, with the full power of the racial unconscious, to preserve itself—and I had no wish to tilt at the windmill.

I confess that my first thought was of simply giving my money away, in a stupendous orgy of charity, and taking a job in a garage. But John was wise enough to be able to show me that that would have been as practical as disposing of a warehouse full of high explosive by setting fire to it with a match. You may have read in newspapers, some years back, of a young man who attempted to give away an inheritance, a *much* smaller fortune than mine. He is now hopelessly insane, shattered by the power that was thrust upon him. *He did not do it to himself.*

So I started small, and very slowly. The first thing I did was to heal the ulcers of the hospital's accounting department. They had been juggling desperately to cover the cost of the care that John Smiley was getting, so I bought the hospital and told them to juggle away, whenever they felt they should. That habit was hard to break; I bought forty-seven hospitals in the next two years, and quietly instructed them to run whatever loss they had to, to provide maximum care and comfort for their patients. I spent the next six years working in them, a month or two each, as a janitor. This helped me to assess their management, replacing entire staffs down to the bedpan level when necessary. It also added considerably to my education. There are many hospitals in the world, Mr. Campbell, some good, some bad, but I know for certain that forty-seven of them are wonderful places in which to hurt.

The janitor habit was hard to break, too. Over the next ten years I toured my empire, like a king traveling incognito to learn the *flavor* of his land. I held many and varied jobs, for my empire is an octopus, but they all amounted to janitor. I spent ten years toiling anonymously at the very borders of my fortune, at the last interface between it and the people it involved, the communities it affected. And without me at the helm, for ten *years*, the nature and operation of my fortune changed in no way whatsoever, and when I realized that it shook me I gave up my tour of inspection and went to my estate in British Columbia and holed up for a few years, thinking it through. Then I began effecting changes. This used-car lot is only one of them. It's my favorite, though, so it's the first one I've implemented and it's where I choose to spend my personal working hours.

But there are many other changes planned.

3.

The silence stretched like a spring, but when at last I spoke my voice was soft, quiet, casual, quite calm.

"And you expect me to believe that none of these changes will make a profit?"

He blinked and started, precisely as if a tape recorder had started talking back to him.

"My dear Mr. Campbell," he said with a trace of sadness, "I frankly don't expect you to believe a word I've said."

My voice was still calm. "Then why tell me all this?"

"I'm not at all sure. But I believe it has much to do with the fact that you are the first person to ask me about it since I opened this shop."

Calm gone. "Bullshit," I roared, much too loud. "Bullfuckingshit, I mean a king-size meadow-muffin! Do you goddammit," I was nearly incoherent, "think I was fucking born yesterday? Sell me a free lunch? You simple sonofabitch *I am not that stupid!*"

This silence did not stretch; it lay there like a bludgeoned dove. I wondered whether all garages echoed like this and I'd never noticed. *The hell with control, I don't need control, control is garbage, it's just me and him.* My spine was very straight.

"I'm sorry," he said at last, as sorrowfully as though my anger were truly his fault. "I humbly apologize, Mr. Campbell. I took you for a different kind of man. But I can see now that you're no fool."

His voice was infinitely sad.

"I don't mind a con, but this is stupid. You're giving away cars and you and Larsen are plenty to handle the traffic. I'm your only customer—what do you take me for?"

"The first wave has passed," he said. "There are only so many fools in any community, only a few na-ive or desperate enough to turn out for a free lunch. It was quite busy here for six months or so, but now all the fools have been accommodated. It will be weeks, months, before word-of-mouth gets around, before people learn that the cars I've sold them are good cars, that my guarantees are genuine. Dozens will have to return, scream for service, promptly receive it and numbly wander home before the news begins to spread. It will get quite busy again then, for a while, and probably very noisy, too—but at the moment I'm not even a Silly Season filler in the local paper. The editor killed it, as any good editor would. He's no fool, either.

"I'm recruiting fools, Mr. Campbell. There was bound to be a lull after the first wave hit. But I believe that the second will be a tsunami."

My voice was a whip. "And this is how you're going to save the world? By doing lube jobs and fixing mufflers?"

"This is one of the ways, yes. It's not surgery, but it should comfort the patient until surgery can be undertaken. It's hard to concentrate on *any*thing when you have a boil on your ass."

"*What?*"

"Sorry. A metaphor I borrowed from John Smiley, at the same time I borrowed the idea itself. 'Ray,' he said to me, 'we're talking about using your money to make folks more comfortable, to remove some of the pointless distractions so they have the energy to sit down and think. Well, the one boil on *every*body's ass is his vehicle—everybody that has to have one, which is most everybody.' Everywhere I went over the next decade, I heard people bitterly complaining about their cars, pouring energy and money into them, losing jobs because of them, going broke because of them, being killed because of them. So I'm lancing the boil, in this area anyway.

"It makes an excellent test-operation, too. If people object too strongly to having their boils lanced, then I'll have to be *extremely* circumspect in approaching their cancers. Time will tell."

"And no one's tried to stop you from giving away cars?"

"I don't give away cars. I sell them at a fair price. But the effect is similar, and yes, there have been several attempts to stop me by various legal means. But there has never been a year of my life when I was being sued for less than a million dollars.

"Then there were the illegal attempts. For a while this lot was heavily, and unobtrusively, guarded, and twice those guards found it necessary to break a few arms. I've dismissed them all for the duration of the lull between waves, but there'll be an army here if and when I need it.

"But until the next wave of customers hits, the only violence I'm expecting is a contract assassination or two."

"Oh?"

The anger drained from my voice as professional control switched in again. I noted that his right hand was out of sight behind his chair—on the side I had not yet seen. I sat bolt upright.

"Yes, the first one is due any time now. He'll probably show up with a plausible identity and an excellent cover-story, and he'll probably demand to see the manager on the obvious pretext. He'll wear strikingly gaudy shoes to draw the attention of casual witnesses from his face, and his shirt will have a high collar and he'll hold his spine very straight. He'll be completely untraceable, expensive, and probably good at his work, but his employers will almost certainly have kept him largely in the dark, and so he'll underestimate his opposition until it is too late. Only then will he realize that I could have come out of that pit with an M-16 as easily as with a pipe wrench if the situation had seemed to warrant it. What *is* that thing, anyway? It's too slim for a blow-gun."

If you've lost any other hope of misdirecting the enemy, try candor. I sighed, relaxed my features in a gesture of surrender, and *very* slowly reached up and over my shoulder. Gripping the handle that nestled against my last few vertebrae, I pulled straight up and out, watching the muscles of his right arm tense where they disappeared behind the chair and wishing mightily that I knew what his hand was doing. I pointedly held the weapon in a virtually useless overhand grip, but I was unsettled to see him pick up on that—he was altogether too alert for my taste. *Hang on, dammit, you can still pull it off if you just hang on.*

"Stiffened piano wire," I said, meeting his eyes, "embedded in a hardwood grip and filed sharp. You put it between the right two ribs and shove. Ruptures the heart, and the pericardial sac self-seals on the way out. Pressure builds. If you do it properly, the victim himself thinks it's a heart attack, and the entry wound is virtually undetectible. A full-scale autopsy would pick it up—but when an overweight car dealer in his fifties has a heart attack, pathologists don't generally get up on their toes."

"Unless he happens to be a multibillionaire," Cardwell noted.

"My employers will regret leaving me in ignorance. Fluoroscope in the fence gate?"

"The same kind they use in airports. If that weapon hadn't been so damned interesting, you'd never have reached the garage."

"I wanted to do the research, but they were paying double for a rush job." I sighed. "I knew better. Or should have. Now what?"

"Now let go of that thing and kick it far away."

I did so at once.

"Now you can have another beer and tell me some things."

"Sorry, Cardwell. No names. They sent me in blind and I'll speak to them about that one day, but I don't give names. It's bad for business. Go ahead and call the man."

"You misunderstand me, sir. I already know Hakluyt's name quite well, and I have no intention of calling police of any description."

I knew the location of every scrap of cover for twenty yards in any direction, and I favored the welding tanks behind me and to my left—he looked alert enough not to shoot at them at such close range, and they were on wheels facing him. If I could tip my chair backwards and come at him from behind the tanks . . .

". . . and I'd rather not kill you unless you force me to, so please unbunch those muscles."

There was no way he was going to let me walk away from this, and there was no way I was going to sit there and let him pot me at his leisure, so there was no question of sitting still, and so no one was more surprised than me when the muscles of my calves and thighs unbunched and I sat still.

Perhaps I believed him.

"Ask your questions," I said.

"Why did you take this job?"

I broke up. "Oh, my God," I whooped, "how did a nice girl like me wind up in such a profession, you mean?" The ancient gag was suddenly very hilarious, and I roared with laughter as I gave the punchline. "Just lucky, I guess."

Pure tension-release, of course. But damned if he didn't laugh at the old chestnut, too—or at himself for all I know. We laughed together until I was done, and then he said, "But why?" and I sobered up.

"For the money, of course."

He shook his head. "I don't believe you."

What's in your right hand, old man? I only shrugged. "It's the truth."

He shook his head again. "Some of your colleagues, perhaps. But I watched your face while I told you my story and *your* empathic faculty seems to be functioning quite nicely. You're personally involved in this, involved with me. You're too damn mad at me and it's confusing you as you sit there, spoiling your judgment. Oh no, son, you can't fool me. You're *some* kind of idealist. But *what brand?*"

There isn't a policeman in the world who knows my name, none of my hits have so much as come to the attention of Homicide, and the reason for it is that my control is flawless, I am an unflappable killing machine, like I said, my emotions aren't even in circuit, and well yes, I had gotten hot under the collar a

couple of times this afternoon for reasons I would
certainly think about when I got a chance, but now of
course it was killingfloor time and I was in total command, and so I was again surprised and shocked to
find myself springing up from my chair and, not div-
ing behind the welding tanks, or even leaping for his
right hand, but simply running flat out full tilt in
plain sight for the door.

It was the most foolish imaginable move and a half
of my mind screamed, *Fool! Fool! At least run broken
field your back is a fucking perfect target you'll never
get halfway to the door* with every step until I was
halfway to the door and then it shut up until I had
reached the door and then the other half said quietly *I
knew he wouldn't shoot* but then I had the door open
and both halves screamed. It hadn't occurred to any
of us that the sound system might be antiquated
enough to use those miserable eight-track tapes.

Eight-tracks break down frequently, they provide
mediocre sound quality under the best playback, their
four-program format often leaves as much as ten minutes of dead air between programs, and you can't re-
wind or cue them. And they don't shut themselves off
when they're done. They repeat indefinitely.

> Hunger stopped him
> He lies still in his cell
> Death has gagged his accusations
> We are free now
> We can kill now
> We can hate now
> Now we can end the world
> We're not guilty
> He was crazy
> And it's been going on for ten thousand years!

It is possible for an unrestrained man to kill himself
with his hands. I moved to do so, and Cardwell hit me

from behind like a bag of cement. One wrist broke as I landed, and he grabbed the other. He shouted things at me, but not loud enough to be heard over the final chorus:

Take your place on the Great Mandala
As it moves through your brief moment of time
Win or lose now: you must choose now
And if you lose you've only wasted your (*life is what it really was even if they called it five years he never came out the front door again so it was life imprisonment, right? and maybe the Cong would've killed him just as dead but they wouldn't have raped him and they wouldn't have starved him not literally we could have been heroes together if only he hadn't been a fucking coward coward coward . . .*)

"Who was a coward?" Cardwell asked distantly, and I took it the wrong way and screamed, "*Him!* Not me! HIM!" and then I realized that the song had ended and it was very very silent out, only the distant murmuring of highway traffic and the power-hum from the speakers and the echo of my words; and I thought about what I had just said, and seven years' worth of the best rationalizations I ever built came thundering down around my ears. The largest chunk came down on my skull and smashed it flat.

Gil, I'm sorry!

4.

Ever since Nam I've been accustomed to coming awake instantly—sometimes with a weapon in my hand. I had forgotten what a luxurious pleasure it can be to let awareness and alertness seep back in at their own pace, to be truly *relaxed*. I lay still for some time, aware of my surroundings only in terms of their peacefulness, before it occurred to me to identify them. Nor did I feel, then, the slightest surprise or alarm at the defection of my subconscious sentries. It was as though in some back corner of my mind a dozen yammering voices had, for the first time within memory, shut up. All decisions were made.

I was in the same chair I'd left so hastily. It was tilted and reshaped into something more closely resembling the acceleration cradles astronauts take off in, only more comfortable. My left wrist was set and efficiently splintered, and hurt surprisingly little. Above me girders played geometric games across the high-curved ceiling, interspersed with diffused-light fixtures that did not hurt to look at. Somewhere to my left, work was being done. It produced sound, but sound is divided into music and noise and somehow this clattering wasn't noise. I waited until it stopped, with infinite patience, in no hurry at all.

When there had been no sound for a while I got up and turned and saw Cardwell again emerging from the pit beneath the Rambler, with a thick streak of grease across his forehead and a skinned knuckle. He beamed. "I love ball joints. Your wrist okay?"

"Yes, thanks."

He came over, turned my chair back into a chair, and sank into his own. He produced cigarettes and gave me one. I noticed a wooden stool, obviously handmade, lying crippled near a workbench. I realized that Cardwell had sawed off and split two of its legs to make the splints on my wrist. The stool was quite old, and all at once I felt more guilt and shame for its destruction than I did for having come to murder its owner. This amused me sourly. I took my cigarette to the front of the garage, where one of the great bay doors now stood open, and watched night sky and listened to crickets and bull frogs while I smoked. Shop closed, Arden gone home. After a while Cardwell got up and came to the door, too, and we stepped out into the darkness. The traffic, too, had mostly gone home for the night, and there was no moon. The dark suited me fine.

"My name," I said softly, "is Bill Maeder."

From out of the black Cardwell's voice was serene. "Pleased to meet you," was all he said.

We walked on.

"I used to be a twin," I said, flicking the cigarette butt beneath my walking feet. "My brother's name was Gil, and we were identical twins. After enough people have called your twin your Other Half, you begin to believe it. I guess we allowed ourselves to become polarized, because that suited everyone's sense of symmetry or some damned thing. Yin and Yang Maeder, they called us. All our lives we disagreed on everything, and we loved each other deeply. Then they called us in for our draft physical. I showed up and he didn't and so they sent me to Nam and Gil to Leavenworth. I walked through the jungles and came out a hero. Gil died in his cell at the end of a protracted hunger strike. A man who is starving to death smells like fresh-baked bread, did you know that? I spent my whole first furlough practically living in his cell, arguing with him and screaming at him, and he just sat there the whole time smelling like whole wheat right out of the oven."

Cardwell said nothing. For a while we kept strolling. Then I stopped in my tracks and said, "For seven years I told myself that *he* was the coward, that he was the chump, that he had failed the final test of survival. My father is a drunk now. My mother is a Guru Maharaj Ji premie." I started walking again, and still Cardwell was silent. "I was the coward, of course. Rather than admit I was wrong to let them make me into a killer, I gloried in it. I went free lance." We had reached my Dodge, and I stopped for the last time by the passenger-side door. "Goodness, sharing, caring about other people, ethics and morals and all that—as long as I believed that they were just a shuck, lies to keep the sheep in line, I could function, my choice made sense. If there is no such thing as hope, despair can be no sin. If there is no truth, one lie is no worse than another. Come to think of it, your Arden said something like that." I sighed. "But I hated that god-damned Mandala song, the one about the draft re-sister who dies in jail. It came out just before I was shipped out to Nam." I reached through the open car window and took the Magnum from the glove compartment. "Right after the funeral." I put the barrel between my teeth and aimed for the roof of my mouth.

Cardwell was near, but he stood stock-still. All he said was, "Some people never learn."

My finger paused on the trigger.

"Gil will be glad to see you. You two tragic expiators will get on just fine. While the rest of us clean up the mess you left behind you. Go ahead. We'll manage."

I let my hand fall. "What are you talking about?"

All at once he was blazing mad, and a multibillionaire's rage is a terrible thing. "You simple egocentric bastard, did it ever occur to you that you might be *needed?* That the brains and skills and talent you've been using to kill strangers, to play head-games with yourself, are scarce resources? Trust an assassin to be arrogant; you colossal jackass, *do you think Arden Larsens grow on trees?* A man in my kind of business can't recruit through the want ads. But I need people with *guts.*"

"To do what?" I said, and threw the pistol into the darkness.

APPENDIX

THIS TIME NEXT YEAR

1) This time next year   I will have won or lost
   This time next year   my bri‑dges all will be crossed
2) This time next year   I will be   up   or   down
   Far away from here   or still hung up   in   town
3) This time next year   I'll ei‑ther win   or   lose
   This time I fear   I'm on a   short, short fuse

1) I'll ei‑ther be in   head‑lines
   Or stand‑in' in the   bread‑lines
2) I'll ei‑ther be in   clo‑ver
   Or bare‑ly turn‑in'   o‑ver
3) I'll ei‑ther be to‑ge‑ther
   En‑joy‑ing   sun‑ny   wea‑ther

1) It all de‑pends on how the dice are   tossed
2) It all de‑pends on how the deal goes down
3) Or suck‑in' up an aw‑ful lot of booze

I   feel it   com‑in' on   Its   oh so   close now

Won‑der   if it's   bad or good   Hope it is‑n't   gon‑na be an

O‑ver dose now   Real‑ly wish   I   knew where I   stood   VERSE 3

COME TO MY BEDSIDE

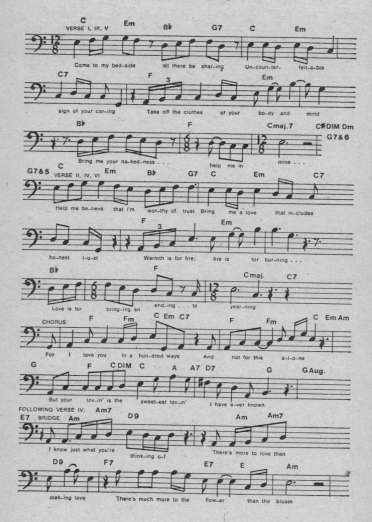

Magnificent Fantasy From Dell

Each of these novels first appeared in the famous magazine of fantasy, *Unknown*—each is recognized as a landmark in the field—and each is illustrated by the acknowledged master of fantasy art, Edd Cartier.

Comes the Blind Fury

John Saul

**Bestselling author of
Cry for the Strangers
and *Suffer the Children***

More than a century ago, a gentle, blind child walked the paths of Paradise Point. Then other children came, teasing and taunting her until she lost her footing on the cliff and plunged into the drowning sea.

Now, 12-year-old Michelle and her family have come to live in that same house—to escape the city pressures, to have a better life.

But the sins of the past do not die. They reach out to embrace the living. Dreams will become nightmares.

Serenity will become terror. There will be no escape.

A Dell Book $2.75   (11428-4)

Dell Bestsellers

At your local bookstore or use this handy coupon for ordering:

THE FAR CALL

by Gordon Dickson

The people and politics behind a most daring
adventure—the manned exploration of Mars!

In the 1990s Jens Wylie, undersecretary
for space, and members of four other nations,
are planning the first manned Mars voyage.
But when disaster hits, it threatens the
lives of the Marsnauts and the destiny of the
whole human race and only Jens Wylie
knows what has to be done!

*A Quantum Science Fiction novel
from Dell $2.25*

World Class SF & Fantasy by the Masters

THEODORE STURGEON
- [ ] The Dreaming Jewels $1.75 (11803-4)
- [ ] Visions and Venturers $1.75 (12648-7)
- [ ] The Stars Are the Styx $2.25 (18006-6)

PHILIP K. DICK
- [ ] The Zap Gun $1.75 (19907-7)
- [ ] Time Out of Joint $2.25 (18860-1)

**L. SPRAGUE DE CAMP &
FLETCHER PRATT**
- [ ] Land of Unreason $1.75 (14736-0)

JACK WILLIAMSON
- [ ] Darker Than You Think $1.75 (11746-1)

EDGAR PANGBORN
- [ ] Still I Persist in Wondering $1.75 (18277-8)

L. RON HUBBARD
- [ ] Slaves of Sleep $1.75 (17646-8)

Quantum Science Fiction

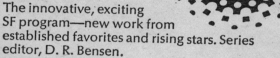

The innovative, exciting SF program—new work from established favorites and rising stars. Series editor, D. R. Bensen.

DREAM SNAKE

Vonda N. McIntyre

"Rich in character, background and incident—unusually absorbing and moving."

Publishers Weekly

"This is an exciting future-dream with real characters, a believable mythos and, what's more important, an excellent readable story."

Frank Herbert

The *"haunting, rich and tender novel"** of a unique healer and her strange ordeal.

* Robert Silverberg

A Dell Book $2.25   (11729-1)